Hybrid Renewable Energy Systems for Remote Telecommunication Stations

Adel A. Elbaset • Salah Ata

Hybrid Renewable Energy Systems for Remote Telecommunication Stations

Springer

Adel A. Elbaset
Electrical Engineering Department
Faculty of Engineering, Minia University
El-Minia, Egypt

EL-Arish High Institute of Engineering
and Technology
EL-Arish, North Sinai, Egypt

Salah Ata
Telecom Egypt
El-Minia, Egypt

ISBN 978-3-030-66343-8 ISBN 978-3-030-66344-5 (eBook)
https://doi.org/10.1007/978-3-030-66344-5

This Springer imprint is published by the registered company Springer Nature Switzerland AG
The registered company address is: Gewerbestrasse 11, 6330 Cham, Switzerland

Preface

The world has become a small village, and that is due to the remarkable scientific advances of communication systems. But there are obstacles to the arrival of communications service to remote and distant locations, and the main problem is feeding these stations with electrical energy, especially in light of the difficulty of connecting these places to the public UG due to the very high economic cost. To supply communications services to remote areas, it is needed to provide an economical electricity source. That is why this book was central to the optimal electrical system that relies on renewable energy to feed communications stations in remote locations.

Currently, in Egypt, there are communications stations on the desert road and in a few very remote places because these stations rely on solar panels as the main source and diesel machine as backup source. Batteries are used to feed loads during the time period required for switching from the main source to backup source and vice versa. The drawbacks of these systems are many like fossil fuels that are needed by the diesel machine also output pollution from the diesel machine and also it would need periodic maintenance. Also, batteries have short life expectancy and contain environmentally harmful chemicals—even batteries are affected by changing temperatures.

The proposed system depends on using solar panels or wind turbines or UG as the main source and fuel cells as a backup source. Ultracapacitors feed electrical loads during the switching time periods that required to switch between main source and backup source; also ultracapacitors cover the sudden change in loads or when there is a fluctuation in the power source. Then the proposed power system can reach 100% reliability so there is no interruption of the electrical power supply to the loads.

Abbreviations

AC	Alternative Current
ADSL	Asymmetric Digital Subscriber Line
AFC	Alkaline FC
AI	Artificial Intelligence
ANN	Artificial Neural Network
AON	Active Optical Network
BS	Battery Storage
BTS	Base Transceiver Station
CCM	Continuous Conduction Mode
CM	Cable Modem
CO	Central Office
CO_2	Carbon Dioxide
COE	Cost of Electricity
CPE	Customer Premises Equipment
CRF	Capital Recovery Factor
CSP	Concentrated Solar Power
DC	Direct Current
DMFC	Direct methanol Fuel cell
DOCSIS	Data over Cable Service Interface Specification
DRFC	Discrete Regenerative Fuel Cell
DSL	Digital Subscriber Line
DSLAM	DSL Access Multiplexer
F	Faraday's constant
FC	Fuel Cell
FTTB	Fiber-To-The-Building
FTTC	Fiber-To-The-Curb
FTTH	Fiber-To-The-Home
FFTC	Fiber-To-The-Curb
FFTN	Fiber-To-The-Node

FTTx	Fiber to the x
EMS	Energy Management System
ESR	Equivalent Series Resistance
EZ	Electrolyzer
GDE	Gas Diffusion Electrode
GPON	Gigabit Passive Optical Network
HOMER	Hybrid Optimization Model for Electric Renewable
HRES	Hybrid Renewable Energy System
HSWPS	Hybrid Solar/Wind Power System
h	Hour
IEEE	Institute of Electrical and Electronics Engineers
IGBT	Insulated-Gate Bipolar Transistor
IC	Initial Capital Cost of Each Component
ICT	Information and Communication Technology
IT	Information Technology
kW	Kilo-Watt
kWh	Kilo-Watt hour
LCA	Life Cycle Assessment
LEC	Levelized Energy Cost
LPSP	Loss of Power Supply Probability
LUCE	Levelized Unit Electricity Cost
MCFC	Molten Carbonate FC
MFC	Microbial Fuel Cell
MLI	Multilevel Inverter
MPP	Maximum Power Point
MPPT	Maximum Power Point Tracker
MSC	Mobile Switching Centers
MW	Mega-Watt
MWh	Mega-Watt hour
N	Avogadro's number
NASA	National Aeronautics and Space Administration
OLT	Optical Line Termination
O&M	Operation and Maintenance Cost
ONWTG	Optimum Number of WTG
OPEX	Operating Exchange
PAFC	Phosphoric Acid Electrolyte Fuel cell
PCFC	Protonic Ceramic Fuel Cell
PEMFC	Polymer Electrolyte Membrane FC
PON	Passive Optical Network
P&O	Perturbation and Observation
PC	Power Conditioning
P-t-P	Point to Point Networks
P-t-MP	Point to Multipoint Networks
PV	Photovoltaic

PVPS	Photovoltaic Power System
PWM	Pulse Width Modulation
RBS	Radio Base Stations
RF	Radio Frequent
RFC	Regenerative Fuel cell
RE	Renewable Energy
REBS	Renewable Energy Base Station
RES	Renewable Energy Sources
RN	Remote Node
SOFC	Solid Oxide Fuel Cell
SPWM	Sinusoidal Pulse Width Modulation
STP	Standard Temperature and Pressure
STC	Standard Test Conditions
Std.	Standard
TCF	Temperature Correction Factor
TPC	Total Present Cost
TWh	Tera Watt hour
UC	Ultracapacitor
UG	Utility Grid
URFC	Unitized Regenerative Fuel Cell
VDSL	Very High Bit-Rate Digital Subscriber Line
VSI	Voltage Source Inverter
WES	Wind Energy System
WTG	Wind Turbine Generator
ZAFC	Zinc-Air Fuel Cell
°C	Celsius (Degrees)
°N	North Direction
°W	West Direction

Symbols

A	P-N junction ideality factor
α_1	Horizon elevation angle (degree)
β	Tilt angle of solar cell modules (degrees)
d	Length of PV module for each array (m)
δ	Declination angle (degrees)
ΔI_L	Peak-to-peak inductor current (A)
$\varnothing$	Site latitude (degrees)
E_{go}	Band gap energy of semiconductor (eV)
f_s	Switching frequency (Hz)
h	Height of wind turbine hub (m)
H	Solar irradiance (kW/m^2)
$\overline{H}$	Average daily radiation on horizontal surfaces
$\overline{H}_d$	Monthly average daily diffuse radiation

$\overline{H}_{o}$	Mean daily extraterrestrial radiation
H_{STC}	Solar irradiance at STC [$H_{STC} = 1$ kW/m^2]
$\overline{H}_{T}$	Average daily radiation on tilted surfaces
$I(t)$	Hourly output current of PV module (A)
I_a	PV array output current (A)
$I_{L,\ avg}$	Average inductor current (A)
I_{mpp}	Current at P_{max} of PV module (V)
$I_{mpp\ \_\ sub}$	Subsystem output voltage/inverter input voltage (A)
$I_o(t)$	Hourly reverse saturation current (A)
I_{or}	Saturation current at T_r
$I_{ph}(t)$	Hourly generated/photocurrent of solar cells module (A)
I_{sc}	Short circuit current of module at STC (A)
$I_{sc\ \_\ a}$	Short circuit current for resultant PV array (A)
IC_X	Initial capital cost ($/kW)
K	Boltzmann constant (1.380658*10^{-23} J/°K)
K_i	Short circuit current temperature coefficient (A/C°)
$\overline{K}_{T}$	Fraction of mean daily extraterrestrial radiation
L_{min}	Minimum boost converter inductance for CCM (Henry)
LT	Component lifetime (year)
n	Recommended average day for each month
N_p	Parallel-connected PV modules
N_{PV}	Total number of PV modules
N_s	Series-connected PV modules
N_{sub}	Total number of subsystem
OMC_X	Operation and maintenance ($/kW)
P_{max}	Nominal DC peak power of module (kWp)
$P_{PV,\ out}(t)$	Hourly output power of PV module (W)
P_{system}	Total power of proposed system (100 kW)
q	Electron charge (1.60217733*10^{-19} °C)
ρ	Albedo or reflected radiation of the Earth's surface
$\overline{R}$	Ratio between radiation on tilted surfaces to radiation on horizontal surfaces
$\overline{R}_b$	Ratio of monthly average beam radiation on tilted surface to that on horizontal surfaces
R_s	Series resistance of the module (Ω)
R_{sh}	Shunt resistance of the module (Ω)
RC_X	Replacement cost ($/kW)
$T(t)$	Cell working temperature (°K)
T_r	Reference temperature of PV cell (°K)
$V(t)$	Hourly output voltage of PV module (V)
V_a	PV array output voltage (V)
v_{dc}	DC-DC boost converter output voltage (V)
V_{DC}	Output voltage of DC-DC boost converter (V)
V_{mpp}	Voltage at P_{max} of PV module (V)

V_{mpp_sub} Subsystem output voltage/inverter input voltage (V)
V_{oc} Open circuit voltage of module (V)
V_{oc_a} Open circuit voltage for resultant PV array (V)

Contents

1 Introduction and Literature Review . . . 1
1.1 Background . . . 1
1.2 Motivation . . . 2
1.3 Problem Definition . . . 3
1.4 Renewable Energy Sources . . . 4
1.4.1 Wind Energy . . . 4
1.4.2 Solar Energy . . . 4
1.4.3 Hydraulics . . . 5
1.4.4 Geothermal Energy . . . 5
1.4.5 Biomass . . . 5
1.5 Review of Related Work . . . 5
1.6 Approach . . . 9
1.7 Book Outlines . . . 9
1.8 Summary . . . 10

2 Analyze the Types of Communication Stations . . . 11
2.1 Cellular Networks . . . 11
2.2 Wired Networks Energy Consumption . . . 14
2.3 Summary . . . 17

3 Regenerative Fuel Cells as a Backup Power Supply . . . 19
3.1 Introduction . . . 19
3.2 Types of Fuel Cells . . . 21
3.2.1 Polymer Electrolyte Membrane FC (PEMFC) . . . 21
3.2.2 Molten Carbonate FC (MCFC) and Solid Oxide FC (SOFC) . . . 22
3.2.3 Phosphoric Acid Electrolyte FC (PAFC) and Alkaline FC (AFC) . . . 22
3.2.4 Direct Methanol FC (DMFC) . . . 22
3.2.5 Alkaline Fuel Cell (AFC) . . . 23
3.2.6 Zinc-Air Fuel Cell (ZAFC) . . . 23
3.2.7 Protonic Ceramic Fuel Cell (PCFC) . . . 23

3.2.8 Microbial Fuel Cell (MFC) 23
3.2.9 Regenerative Fuel Cells 24
3.3 Principles of Regenerative Fuel Cells 24
3.3.1 Discrete Regenerative Fuel Cells (DRFC) 25
3.3.2 Unitized Regenerative Fuel Cells 25
3.4 Efficiency of Regenerative Fuel Cell 26
3.5 Methodology and Emphasis 27
3.5.1 Hydrogen Production Estimation 27
3.5.2 Oxygen Production Estimation 29
3.5.3 Hydrogen Usage in the Fuel Cell Part 29
3.5.4 Air Usage in Fuel Cell Part 30
3.5.5 Air Exit Flow Rate 30
3.6 Hydrogen Storage 31
3.7 Materials Balance in DRFC 31
3.7.1 Hydrogen Balance 32
3.7.2 Water Balance 32
3.7.3 Oxygen Balance 32
3.8 Summary 33

4 Optimum Sizing of Ultracapacitors 35
4.1 Introduction 35
4.2 Ultracapacitor Characteristics 37
4.2.1 Temperature Dependency 38
4.2.2 Voltage Dependency 38
4.3 Ultracapacitor Electrical Models 40
4.3.1 RC Serial Model 40
4.3.2 First-Order Model for UC 41
4.4 Methodology of UC Sizing 41
4.4.1 Simple Sizing of UC 42
4.4.2 Accurate Sizing 44
4.4.3 UC Sizing Using Artificial Neural Networks 48
4.5 Case Study and Applications 48
4.5.1 Results of Simple Sizing 48
4.5.2 Results of Accurate Sizing 49
4.5.3 Comparative Study Between Simple Sizing and Accurate Sizing of UC 50
4.5.4 ANN Results for Detecting UC Suitable Type 52
4.6 Summary 59

5 Design and Sizing of Photovoltaic Power Systems 61
5.1 Introduction 61
5.2 Estimation of Total Solar Radiation on an Inclined Surface 65
5.2.1 Estimation of Daily Beam Radiation ($\overline{H}_B$) 65
5.2.2 Estimation of Ground Reflection Radiation ($\overline{H}_R$) 65
5.2.3 Estimation of Diffused Sky Radiation ($\overline{H}_S$) 66

5.3 Maximum Power Point Tracking (MPPT), in Photovoltaic Power System . . . 66
5.3.1 Optimum Solar cells Area (OSCA) Estimation for PVPS . . . 67
5.3.2 Final Solar Cells Area Required . . . 68
5.4 Maximum Power Point Tracking (MPPT) . . . 68
5.4.1 MPPT Algorithms Classification . . . 68
5.5 On-Grid Photovoltaic Power Systems . . . 70
5.5.1 DC-DC Converters . . . 71
5.5.2 Inverters . . . 78
5.6 General Control Grid-Connected PV Systems . . . 80
5.7 Results and Discussions . . . 81
5.7.1 Estimation of Hourly Radiation on Tilted Surfaces . . . 81
5.7.2 Estimation of the Optimum Number of Solar Panels . . . 81
5.7.3 Rates of Hydrogen Production and Usage in PVPS . . . 88
5.7.4 Final Solar Cells Area, FSCA, and the Number of Solar Panels . . . 95
5.7.5 ANN Results for Determining the Duty Cycle . . . 104
5.8 Summary . . . 112

6 Design and Sizing Wind Energy System . . . 115
6.1 Introduction . . . 115
6.2 Wind Turbines . . . 115
6.2.1 Basic Components of a Wind Turbine System . . . 116
6.2.2 Modeling of Wind Turbines . . . 117
6.3 Wind Parameter . . . 123
6.3.1 Wind Speed . . . 123
6.3.2 Wind Classes . . . 123
6.3.3 Wind Shear . . . 124
6.3.4 Weibull Distribution . . . 125
6.3.5 Wind Turbine Power Output . . . 126
6.4 Average Power Output and Average Number of Wind Turbine Generators . . . 128
6.5 Design Algorithms . . . 129
6.5.1 Upgrading Wind Velocities to Hub Height . . . 129
6.5.2 Evolution of Weibull Parameter (*C* and *K*) . . . 129
6.5.3 Evolution of the Capacity Factor, CF, and ANWTG . . . 129
6.5.4 Energy Balance Study and the Optimum Number of WTG . . . 130
6.6 Stand-Alone (Off-Grid) Wind Power System . . . 132
6.7 On-Grid Wind Energy Systems . . . 132
6.7.1 AC/DC Converters (Rectifiers) . . . 133

6.8 Results and Discussions 136
6.8.1 ANN Results for Determining Duty Cycle 151
6.9 Summary 157

7 Optimum Sizing and Design of Renewable Energy System 159
7.1 Introduction 159
7.2 Hybrid System Description 162
7.2.1 The Load Demand Model 162
7.2.2 Solar Panel 162
7.2.3 Wind Turbine 162
7.2.4 Regenerative Fuel Cell 163
7.2.5 Ultracapacitor 163
7.2.6 Converters and Controllers 163
7.2.7 Energy Management and Operation Strategy 163
7.3 System Cost Analysis 164
7.3.1 Problem Statement 164
7.3.2 Cost Estimation 164
7.4 Methodology and Constraints 167
7.4.1 PSO Optimization Algorithm 167
7.5 Results and Discussions 169
7.5.1 Analytic Methods 169
7.5.2 PSO Results and Discussion 172
7.6 Summary 178

8 Conclusions and Recommendations for Future Work 179
8.1 Discussions and Conclusions 179

References 181

Index 191

Chapter 1
Introduction and Literature Review

1.1 Background

The progress in communication networks has changed the way people live, work, and play. Since many people around the world are connected by communication networks, the challenge to provide reliable and cost-effective power solutions to these expanding networks is indispensable for telecom operators [1]. In remote areas, grid electricity is not available or is available in limited quantities. Diesel generators with the backup battery were used for powering these sites. These systems usually located in areas with difficult accessibilities require regular maintenance and are characterized by their high fuel consumption and high transportation cost. Also, due to the rapid depletion of fossil fuel reserves and the increasing demand for clean energy technologies to reduce the greenhouse gas emission (CO_2, NO_X, and SO_X) urgent search for alternative solutions for powering these sites is needed. Thus, stand-alone renewable sources can be a feasible solution for powering these sites [2].

Due to the development of new convergent broadband technologies that comprise voice, video, and data in a single platform system, the telecom operator has grown an interest in reliable power services to provide a guarantee of the quality of service for the remote BTS operation hours and traffic management [3].

The stand-alone system comprising only PV and battery and only wind and battery could not provide reliable power supply at these remote stations due to high traffic demand by these enhanced features. For instance, the power generated by wind turbine and PV array is high and low in the winter seasons since the wind speed and solar radiation are generally high and low respectively at this season. Moreover, during nights, solar energy cannot be utilized while wind energy can be used [4]. With these ideas, both renewable sources should be integrated to form a hybrid solar and wind energy system to meet the load demand. The combination of solar and wind energy, moreover, will result in a substantial reduction in the number of solar panels and the size of the battery and therefore the total cost of the system. In

A. A. Elbaset, S. Ata, *Hybrid Renewable Energy Systems for Remote Telecommunication Stations*, https://doi.org/10.1007/978-3-030-66344-5_1

the present day, the cost of wind energy installation is much lower than that of solar panel installation and the adaptation of hybrid energy system results in a reduction of the battery capital and maintenance cost [5].

1.2 Motivation

Energy is directly related to the most critical economic and social issues which affect sustainable development such as water supply, sanitation, mobility, food production, environmental quality, education, job creation security and peace in a regional and global context. Indeed, the magnitude of change needed is immense, fundamental and directly related to the energy produced and consumed nationally and internationally. Also, it is estimated that almost two billion people worldwide lack access to modern energy resources [6]. Current approaches to energy are non-sustainable and non-renewable. Today, the world's energy supply is largely based on fossil fuels and nuclear power. These sources of energy will not last forever and have proven to be contributors to our environmental problems. In less than three centuries since the industrial revolution, humanity has already burned roughly half of the fossil fuels that accumulated under the earth's surface over hundreds of millions of years. Nuclear power is also based on a limited resource (uranium), and the use of nuclear power creates such incalculable risks that nuclear power plants cannot be insured. After 50 years of intensive research, no single safe long-term disposal site for radioactive waste has been found. Although some of the fossil energy resources might last a little longer than predicted, especially if additional reserves are discovered, the main problem of "scarcity" will remain, and this represents the most significant challenge to humanity [7].

The information and communication technology (ICT) is a potentially significant impact sector in the world competition with consuming energy and changing CO_2 emission and the world's climate. ICT sector can be subdivided into the following information technology (IT), telecommunication technology, and network technologies. The ICT sector is getting consumption energy in percentage which varies between 3% over the total global energy consumption [7]. On the other hand, the ICT sector is emitting 2% of the global CO_2 emissions. Both growth rates are still increasing in the future which have a significant impact on energy demand with worldwide warming and bad changes to the Environmental.

The primary sources of greenhouse gas and carbon-based emissions are directly proportional to energy production and consumption, transport, land-use changes, waste management, etc. The primary causes of changing climate are coming from the proliferation of user devices, which all of them need power and generating heat radiation. So energy consumption and reduction of CO_2 is a crucial driver issue in the new ICT sectors.

Telecommunication is a more dynamic and significant part of the ICT sector. The massive amount of energy is consumed by the telecommunication sector, which is increasing as advantages in information technology are being made. Energy used in the telecommunication sector is estimated for a fixed telecom network, cellular wireless communication, and data center. The telecommunication sector consumes energy approximately 1% through the global energy consumption [8], and on the other hand, this sector is responsible for about less than 1% of the total CO_2 emission in the world. This percentage rate is still increasing in the future [9].

The wireless network can be viewed into three major subsection-core networks that are working as a switching system, interface to fixed network and billings system. The second base station (BTS) is established the radio frequency interface between the network and mobile station. Other, mobile station, which is used by the subscriber making a phone and data call [9]. The core network and radio access network–(base station, MS) is using high energy. It is estimated that over 90% of the energy consumption in a wireless network is due to those two elements [8].

1.3 Problem Definition

The rate of mobile subscribers in the telecommunication sector is rapidly increasing on the demand for transferring information. Nowadays there is approximately 3.7 billion around the world [8], and within a few years, this amount of mobile subscriber will be expected about 6 billion around the world. The operator and subscriber are demanding a great use of energy and on the other hand they are taking the responsibility of a large amount of CO_2 emission into the global environment. The reduction of power consumption and CO_2 emission is being a main issue because of giving some reason. The cost of energy is increasing rapidly in the future, and it is making an undesirable part of the operator cost of the telecommunication sectors. The CO_2 gases emission is also rapidly growing due to the telecommunication sector for using fossil fuel, which has a high impact on global environments. It has become our responsibility to reduce the energy consumption and CO_2 gases emission in the wireless network and telecommunication. Approximately 1% in per cent through the global energy consumption. Also, this sector is responsible for about less than 1% of the total CO_2 emission in the world. This percentage rate is still increasing in the future.

Also, the standby energy system will be longer than the traditional standby energy system that used now besides low maintenance rate, high reliability, and little fault rate.

Thousands of communication stations in the world depend on UG as main energy supply, diesel engines, and batteries as standby. There is a lot of problems from this traditional system as it depends on energy coming from fossils fuels, CO_2 emissions, Diesel fuel needs each a certain period, high operation cost, standby energy system hasn't long life and a lot of maintenance needs.

Fossil fueled conventional power plants produce a large part of human-made CO_2 emissions to the atmosphere; such emissions cause global warming and contribute to the negative environmental and health issues [9]. The poor net efficiency of the existing grid infrastructure and climate changes tend to increase the share of distributed renewable energy sources and to improve the efficiency of the traditional grid infrastructure. An estimate shows that the cost of improving the efficiency of the power system is less than the investments in new power generation plants and the cost of the abatement of greenhouse gas emissions [10]. A new future grid concept also known as Smart Grid, improves the efficiency of the power system, contributes to the reduction of the global warming, increases the transmission and distribution lines capacities, offers ancillary services support and facilitates the utilities ability to supply power during peak load hours or when there are faults in the grid lines [11, 12].

1.4 Renewable Energy Sources

In our time scale, renewable energies are those continuously provided by nature. They come from solar radiation, the core of the Earth, and the gravitational forces of the Moon and the Sun with oceans. There are several types of renewable energies: wind power, solar power, biomass, hydraulics, and geothermic [13].

1.4.1 Wind Energy

The wind energy stands out to be one of the most promising new sources of electrical power in the near term. Many countries promote wind power technology through national programs and market incentives.

Wind power can be generated in distributed wind power plants of up to 5 MW capacity each, or large wind parks interconnecting tens or even hundreds of such plants. There are onshore and offshore wind parks, built into the sea where it is not deeper than 40 m [14]. Wind power is typically fluctuating and cannot be delivered on-demand. Wind power is stored for some seconds in the rotating mass of the wind turbines or as chemical or mechanical energy in batteries or large pump storage systems [15].

1.4.2 Solar Energy

The projected lifespan of the Sun is 5 billion years, which makes it in our time scale an inexhaustible and thus renewable energy. The total energy received at the surface of the Earth is 720 million TWh/year, i.e., 6000 times the current primary consumption of all human activities. But the availability of this energy depends on the day-night cycle, on the latitude of where this energy is captured, on the seasons and the cloud cover [16].

Photovoltaic systems are typically used for distributed or remote power systems with or without connection to the utility grid. Their capacity ranges from a few Watts to several MW. Batteries are usually applied in smaller decentralised supply systems to store solar energy over the night. There are also scenarios for huge PV systems up to 1.5 GW each to be built in desert areas. The electricity yield of PV systems is modeled as a function of the global irradiance on a surface tilted at the respective latitude angle. PV cannot offer any secured capacity [17].

1.4.3 Hydraulics

Hydraulics is currently the first exploited renewable source of electricity. The global potential could, however, be better exploited. Large hydraulics (with a power higher than 10 MW) are exploited almost at the maximum of their potential in industrialized countries [18]. Small hydraulics (of a power lower than 10 MW) are partly made from run-of-river power plants, which depend highly on the river flow rate. The small power plants are quite interesting for decentralized production [16].

1.4.4 Geothermal Energy

The temperature of our planet increases considerably as we get closer to the center. In some zones of our planet, we can find, at depth, water at a high temperature. High-temperature (150–300 °C) that can be pumped towards the surface, producing vapor via exchanges, then turning this vapor as in conventional thermal power stations and producing electricity [19].

1.4.5 Biomass

Biomass is plant or animal material used for energy production (electricity or heat), or in various industrial processes as raw material for a range of products [20]. It can be purposely grown energy crops, wood or forest residues, waste from food crops (wheat straw, bagasse), horticulture (yard waste), food processing (corn cobs), animal farming (manure, rich in nitrogen and phosphorus), or human waste from sewage plants [21].

1.5 Review of Related Work

A large number of international studies have been conducted to study how to power communication stations or microgrid using renewable energy system, some of these studies are:

Jansen et al. [22] presented the powering of telecommunication base stations by using an autonomous renewable energy microgrid, using solar photovoltaic and wind turbine to generate electricity and a regenerative hydrogen fuel cell as backup power for up to 10 days. The system is validated using MATLAB/Simulink software and real-life weather data and optimized for a 25 kW microgrid near Dakar, Senegal.

R. Kaur, et al. (2018) [23] studied new telecommunication towers that are installed in remote/rural areas. With concerns over environmental issues, such towers are to be environmentally friendly. Wind–photovoltaic (PV) based DC microgrid is proposed for supplying power to telecommunication towers in remote/rural areas ensuring reliable, economical, and green power supply. Therefore, techno-economic analysis is carried out to determine the feasibility and cost of electricity (COE) per unit of the proposed DC microgrid.

Rahman [24] evaluates the opportunity of fuel cells combined with photovoltaic (PV) array and battery for the more extended backup period (up to 30 h) for the cellular BTS. A dynamic Energy Management System (EMS) is also proposed to ensure optimal power management of the BTS during both regular and emergency periods. The feasibility of the proposed system is verified numerically by utilizing MATLAB/Simulink.

Altinoz [25] introduces the combined energy systems which include solar, wind, and battery with their grid integration. DC and AC loads are implemented in the designed system. The solar system, wind system, battery management system, and converters are modeled in Matlab/Simulink individually.

Cordiner et al. [26] presents the off-grid radio base stations powered with fuel cells and locally available renewable energy sources. Fuel cells have been integrated as a programmable power generator with a photovoltaic system and an energy storage system. Both electrochemical batteries and hydrogen produced locally with an electrolyzer have been tested as energy storage solutions.

Elbaset and Hassan [27] presents an optimum design of rooftop grid-connected photovoltaic (PV) system installation on an institutional building at Minia University, Egypt.

Belkhiri and Chaker [28] proposed hybrid PV/wind energy and use a battery bank to find the margin of safety which corresponds to the desired reliability with minimal cost for the different sites.

Devabakthuni [29] designed a grid-connected wind and photovoltaic system. A new model of converter control was designed which maintains the voltage of the bus to the grid as constant when the combined system of solar and wind is connected to the AC bus. The model is designed to track maximum power at each point irrespective of changes in irradiance, temperature, and wind speed which affects the power supplied to the grid.

Shen [30] proposed some configurations of hybrid energy harvesting systems, including wind-wind-storage DC power system with BOOST converters, solar-solar storage DC power system with cascade BOOST converters, wind-solar-storage DC power system with BOOST converter and cascade BOOST converter, and wind-solar DC power system with SEPIC converter and BOOST converter. The models of all kinds of systems are built-in Matlab/Simulink and the mathematical state-space models of combined renewable energy systems are also established.

Margaret Amutha and Rajini [31] aimed to investigate the economic, technical, and environmental performance of various hybrid power systems for powering remote telecom. Simulations using the Hybrid Optimization Model for Electric Renewable (HOMER) software are performed to determine the Initial Capital, the Total Net Present Cost (TNPC), the Cost of Energy (COE) as well as the system capacity shortage of the different supply options. The simulation results suggest a suitable hybrid system which would be the feasible solution for the generation of electric power for remote telecom.

Sharma et al. [32] presented modeling and simulation of solar cell/wind turbine/fuel cell hybrid power system that developed using a novel topology to complement each other and to alleviate the effects of environmental variations. The wind power and photovoltaic (PV) power generation techniques and the MPPT of each technique. Then, a new stand-alone wind–PV hybrid generation system is proposed for application to remote and isolated areas.

Salih et al. [33] focused on the optimum size and design of a hybrid power system for powering remote Base Transceiver Station (BTS) sites that are based on the target of minimizing capital and operation costs of system components without compensation of meeting the load demand. Three different system configurations are assessed and compared according to the system's efficiency and performance, Cost of Energy (COE), and environmental emissions. This analysis is carried out by HOMER software.

Noguera [34] studied that remote cell towers need on-site power generation or stand-alone power systems for operation and, traditionally, diesel-electric generators have been used as backup power systems in combination with renewable energies. The first stage of this research involves the design of two stand-alone power systems using a microgrid design software (HOMER), utilizing the mentioned technologies for such applications, in two different scenarios that differ in solar radiation. The purpose is to obtain the components used by the power systems and carry out a comparative environmental analysis, which would consist of a Life Cycle Assessment (LCA). The results of the LCA conclude the use of PEMFC as backup systems.

Olatomiwa et al. [35] illustrated the size optimization of the solar-wind-diesel generator-battery hybrid system designed for a remote location mobile telecom base transceiver station in Nigeria. Different energy combinations have been analyzed using HOMER 2.81 (Hybrid Optimization Model for Electric Renewables) in order to determine an optimal model. Simulation results show that the hybrid energy systems can minimize the power generation cost significantly and can decrease CO_2 emissions as compared to the traditional diesel generator only.

Kusakana and Vermaak [36] investigated the possibility of using hybrid photovoltaic–wind renewable systems as primary sources of energy to supply mobile telephone base transceiver stations in the rural regions of the Democratic Republic of Congo. Four different possible options including a hybrid photovoltaic–wind, a diesel generator, a pure photovoltaic, and a pure wind energy system were designed to compare and evaluate their technical performance, economics, and environmental impact. Simulations using HOMER are performed to determine the initial capital, the total net present cost, the cost of energy as well as the system capacity shortage of the different supply options. The selection criteria include the financial viability, fuel consumption, and CO_2 emissions for a project lifetime of 20 years.

Natsheh [37] presented a novel adaptive scheme for energy management in stand-alone hybrid power systems. The proposed management system is designed to manage the power flow between the hybrid power system and energy storage elements in order to satisfy the load requirements based on artificial neural network (ANN) and fuzzy logic controllers.

Hassan et al. [38] discussed the fundamental principles and motivations necessary for a thorough understanding of the usage of renewable energy in cellular networks. Furthermore, it introduces a reference model for renewable energy base stations (REBS) and provides an analysis of its components. The synthesis of dedicated REBS algorithms, system architectures, deployment strategies, and evaluation metrics provides a comprehensive overview and helps identify the perspectives in this promising research field.

Zhai [39] analyzed the technological feasibility of solar PV, wind, and fuel cells, also; capacity factor, levelized energy cost, cost-benefit, and sensitivity analysis are conducted with different geographic and economic parameters.

Al-Sharafi [40] presents a model for the sizing optimization of a hybrid PV/wind power generation system. The model involves a PV model, wind power model, and a model for the required battery. The developed model has been used to optimize the hybrid PV/wind system for an off-grid house in the eastern province of the Kingdom of Saudi Arabia.

Paudel et al. [41] dealt with the problem of a feasibility assessment and optimum size of photovoltaic (PV) array, wind turbine and battery bank for a stand-alone Hybrid Solar/Wind Power system (HSWPS) at remote telecom station of Nepal. In this work, feasibility analysis is carried out through hybrid optimization model for electric renewables (HOMER) and mathematical models were implemented in the MATLAB environment to perform the optimal configuration for a given load and the desired loss of power supply probability (LPSP) from a set of system components with the lowest value of cost function defined in terms of reliability and levelized unit electricity cost (LUCE).

Nema et al. [42] gave the design idea of optimized PV-solar and wind hybrid energy system for GSM/CDMA type mobile base stations over conventional diesel generators for a particular site in central India (Bhopal). For this hybrid system, the meteorological data of solar insolation, hourly wind speed, are taken for Bhopal-Central India and the pattern of load consumption of mobile base stations is studied and suitably modeled for optimization of the hybrid energy system using HOMER software.

Elbaset and El-Tamaly [43] proposed a computer program to determine the optimal design of the PV system. The proposed computer program based on the minimization of energy purchased from the grid. A comparative study between three different configurations (stand-alone Photovoltaic Power System (PVPS) with

Battery Storage (BS), PVPS interconnected with UG without BS, and grid-connected PVPS accompanied with BS) has been carried out from economic and reliability points of view with the main goal of selecting suitable one, to be installed at Zâfarana site to feed the load requirement.

1.6 Approach

This book is to investigate renewable energy systems that can be generally fed all communication stations found in populated areas or remote areas (rural areas) with using renewable energy or utility grid as a main energy supply and fuel cells and ultracapacitor as standby supply to overcome CO_2 Fossil fuel needs high operation cost, standby energy system hasn't long life and a lot of maintenance needs.

1.7 Book Outlines

To achieve the above objectives, this book is organized in eight chapters, each chapter of the book addresses a specific topic. Main points of each chapter are highlighted as follows:

Chapter 1 contains a brief explanation of the problem to be addressed, as well as a summary of some previous studies on the topic under consideration. This section also deals with a brief description of some types of renewable energies.

Chapter 2 analyzes and displays types of communication stations; the rate of consumption of electrical power by communication stations has also been addressed.

Chapter 3 presents the principles of fuel cells, the idea of their functioning and their internal components—the benefits of fuel use in communications. Regenerative fuel cells (RFC) have been selected in this study. Hydrogen and oxygen output and their relationships to the energy consumed are explained. The rates of used hydrogen for electricity production were also explained. Calculations of the water and hydrogen tanks and ways of hydrogen storage are displayed. How to get RFC efficiency was explained.

Chapter 4 exposes ultracapacitors composition, working theory and characteristics. Also, advantages, the equivalent circuits, temperature effect, and voltage variation effect of ultracapacitors are displayed. This chapter also addressed the methods of ultracapacitors sizing. ANN technique is used to determine the optimum UC type for a certain load. Application and results of applied sizing methods and ANN techniques on different loads are displayed.

Chapter 5 looks into the photovoltaic power system; the calculation of solar radiation incident on surfaces is provided, the required number of solar panels to feed a certain load. Three locations in Egypt are selected to apply the case study, four types of modern solar panels are used to determine the appropriate one. A computer program is prepared to calculate the number of panels required to supply the load in two cases: 1- tilt angle of solar panel fixed that equals latitude, 2- tilt angle of solar panels varies each month. Through what has been done, the best type of solar panels and the optimal number of each location have been identified. A study was also conducted on the connection of the photovoltaic power system to the UG. ANN technique is used to obtain MPPT.

Chapter 6 explores the wind energy power system. This system contains wind turbines as the main source and RFC as a backup source and means of store energy in the form of hydrogen. Equations have been provided for the calculation of generated wind energy. Eleven types of wind turbines with appropriate capacity rates for the loads were selected to choose the best of them according to wind speeds for each of the sites selected for the study also according to the value of the power coefficient of each site, where it is choosing the highest power coefficient because this means that it is the most appropriate for this site. A capacity balance program has also been developed to determine the optimal number of wind turbines required to feed loads, taking into account the efficiency of RFC. The value of hydrogen produced by energy conversion was determined in the surplus times. The value of hydrogen consumption during defect times. The connection of the wind energy system to the UG has also been studied. ANN technique is used to obtain maximum power.

Chapter 7 proposes a powerful system of hybrid (solar panels, wind turbines, regenerative fuel cells, and ultracapacitor). The economic study of the hybrid system is carried out in two ways: analytical method and artificial intelligence method; it has appeared that the final results of the two methods are almost identical. The results showed that there are locations where solar panels are more economically used than air turbines and vice versa. This result is due to the small electrical load, the mixed system of solar panels and wind turbines is costly.

Chapter 8 summarizes the conclusions of the book, recommendations, and future work.

1.8 Summary

A brief explanation of the book problem is addressed, as well as a summary of some previous studies on the topic under consideration. A brief dissection about some types of renewable energies is introduced.

Chapter 2
Analyze the Types of Communication Stations

This chapter provides an overview of the different types of communication networks and stations. Generally, there are mainly two types of communication networks: cellular networks and wired networks. Each network has its infrastructure. For each entity, there will be a short description, and it will be provided the corresponding energy consumption model.

2.1 Cellular Networks

Figure 2.1 shows a simplified representation of a wireless cellular system. It can be considered that consists of three different sectors [44]:

- **Mobile Switching Centers (MSCs)**, which are responsible for switching functions and have the role of the interface to the fixed (core) network.
- **Radio Base Stations (RBSs)**, which represent the access network and offer wireless communication link between mobile terminals and the core of the network.
- **Mobile terminals**, which are the end user's part, usually represented by handheld devices.

Based on energy measurement models, it has been calculated that over 90% of the wireless network power consumption is part of the operator's operating exchange (OPEX), which includes the Mobile Switching Centers (MSCs) and Radio Base Stations (RBSs).

The key elements are the highly deployed radio base stations because the total number of base stations is exceptionally high. There are approximately four million installed Base Transceiver Station (BTS) cabinets in the world today with relatively high energy consumption, about 60 TWh per year. Furthermore, the corresponding CO_2 emissions are around 30 Mt. In general terms, access to radio equipment is responsible for about 80% of the energy consumed by a cellular network [47].

A. A. Elbaset, S. Ata, *Hybrid Renewable Energy Systems for Remote Telecommunication Stations*, https://doi.org/10.1007/978-3-030-66344-5_2

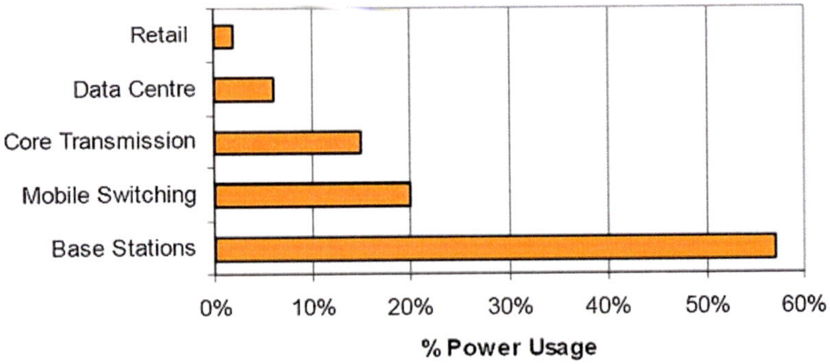

Fig. 2.1 Wireless cellular network power consumption

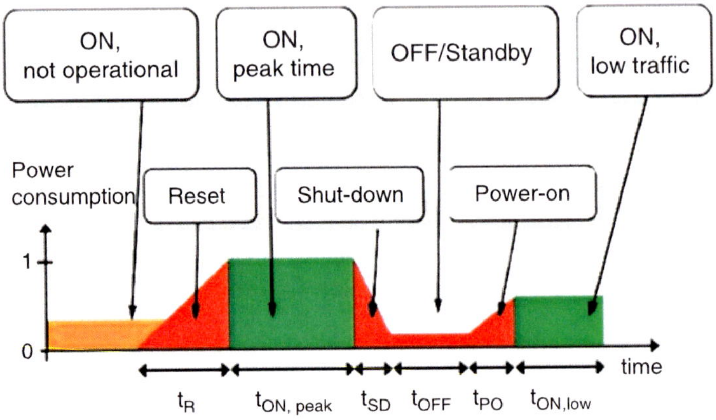

Fig. 2.2 Time-varying behavior of a base station

According to Fig. 2.1, it is evident that base stations are the main energy consumers of a common cellular network. In comparison to the rest of the cellular network elements, they present the highest power consumption rates since a typical radio base station consumes from 800 W to 3 KW [48] and as we highlighted before they represent around 60–80% of the total network energy consumption. This fact leads us to the conclusion that in order to achieve efficient energy management we should mainly concentrate on the base station's design architectures and operation modes.

According to Fig. 2.2, a typical base station is possible to be in an operational or in a non-operational state (idle mode). Additionally, in an operational state, it can experience different traffic load conditions such as peak traffic or low traffic states. As we can notice the power consumption varies over time depending on the different base station modes. Regarding the base station architecture, Fig. 2.3 illustrates the several power-consuming elements which are included within a typical base station cabinet. Firstly, we will provide a short description of each component. Secondly,

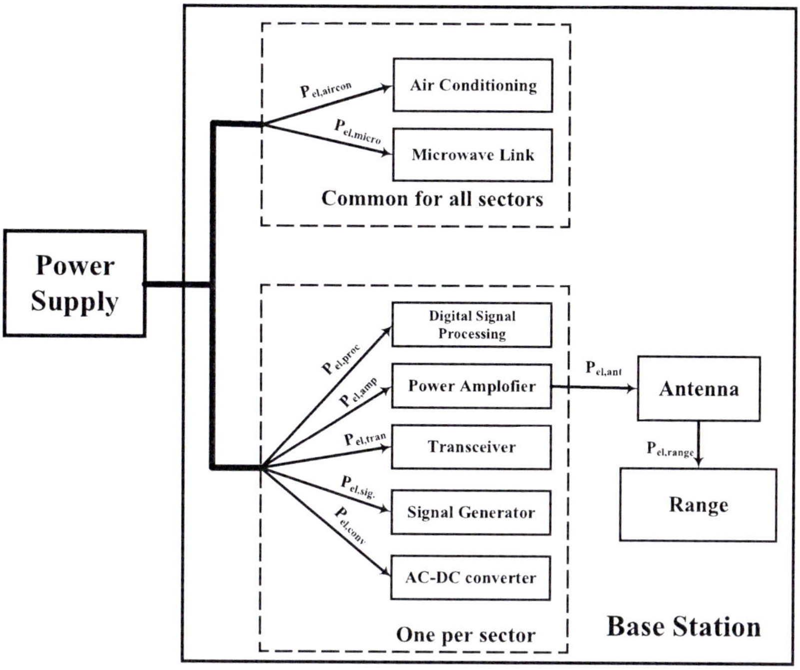

Fig. 2.3 Main components of a base station structure

we will present the relative power amounts consumed by each base station element as were measured in recent researches.

Proceeding to the decomposition process of the base station architecture as it described in [49], Figure 2.3 shows the detailed elements listed below:

- *Air Conditioning*: controls the base station temperature
- *Microwave link*: plays the role of the connector with the core network
- *Digital Signal Processing*: is responsible for the conversion of the analog signal to *a* sequence of bits or symbols
- *Power Amplifier:* main concern of power amplifier is to convert the DC input power into radio-frequent (RF) wave
- *Transceiver:* can receive and transmit the signals
- *AC-DC Converter (Rectifier):* converts alternating current (AC) to direct current (DC)

Generally, the power consumption of a base station is time-varying and it consists of two parts [50]. The first part is related to the fixed (static) power consumption which represents the energy required so that the base station can operate regardless of the presence or not of traffic load. The second part depends on the traffic load conditions and presents dynamic energy consumption. The dynamic part then is added to the fixed part.

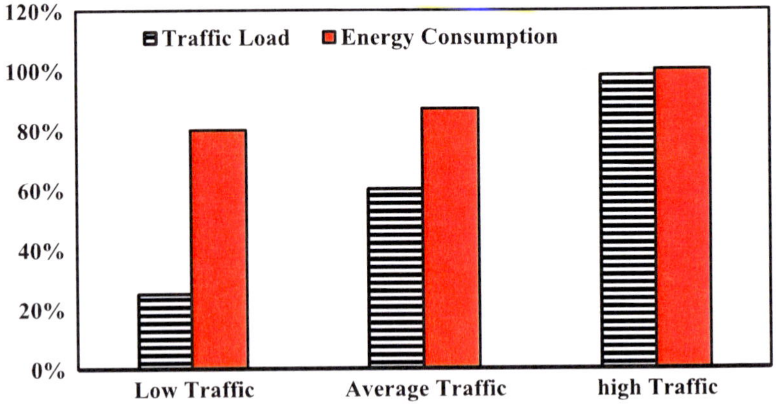

Fig. 2.4 Power consumption rate vs. traffic load rate for base station

For our convenience, we will study the power consumption model of two different types of base stations which is strongly related to area coverage that they provide:

- *Macrocell base stations* offer extensive area coverage, and they can be sectored to smaller micro base stations. For these types of base stations, the contribution of the abovementioned dynamic power part is negligible [50] since it is considered that the total power consumption is almost independent of the instantaneous traffic load for macro sites and it is profoundly affected by static part.
- *Microcell base stations* in design size are more compact and cover small but densely populated areas (microsites). The power consumption here has a more dynamic nature than in macrocell base stations, and it depends on the current user activity level.

Figure 2.4 displays power consumption rate vs. different traffic load conditions (average, low, and high); the base station power consumption in general terms remains at the same levels. For example, it can be seen that at a 75% reduction in traffic load (low traffic), it is not accompanied by a relative decrease in power consumption. Therefore, it is no stable relationship between energy consumption and the traffic load [52].

2.2 Wired Networks Energy Consumption

Having completed the discussion about wireless networks, we now focus our attention on wired communication networks to investigate how the energy is distributed over several power-consuming domains. The following analysis is based on information available in [51].

Table 2.1 Power consumption in wired communication networks [58]

Network domain	Power consumption (W)
Home	10
Access	1280
Metro	6000
Core	10,000

During the year 2002, the fixed broadband networks were responsible for 3% of total CO_2 emissions of the telecommunication industry, and it is expected to increase by up to 14% in 2020. Compared to wireless networks, wired communication networks introduce a different power consumption pattern. With regard to the mobile networks, we have highlighted that the end users represent only 10% of the overall power consumption and for the rest of 90% responsible is the operator's contribution. On the contrary, for the fixed networks, the dominant power consumer is the user's segment representing 70% of the total consumption, and only 30% is due to the operator OPEX [53]. Wired communication networks can be divided into three segments: core, metro, and access as can be shown in Fig. 2.4 [54]. Below, we provide a short description of each domain.

- *Core Network:* By core network, we usually refer to the backbone infrastructure of a telecom network. The core network is typically based on a mesh interconnection pattern and carries enormous amounts of traffic collected through the peripheral areas of the network. So, it needs to be equipped with appropriate interfaces towards the metro and access networks that are in charge to collect and distribute traffic, so that users separated by long distances can communicate with one another through the core (backbone) network [55].
- *Metro Network:* The metro network is the part of a telecom network that typically covers metropolitan regions. It connects equipment for aggregation of residential subscribers' traffic, and it provides direct connections to the core network for internet connectivity [54].
- *Wired Access Network:* The access network is the "last mile" of a telecom network connecting the telecom CO (Central Office) with end users. Access network comprises the larger part of the telecom network. It is also a major consumer of energy due to the presence of a massive number of active elements. There are several network access technologies proposed such as xDSL (Digital Subscriber Line), CM (Cable Modem), FTTx (Fiber-To-The-x), etc. [54, 56].

The scope of the above-described infrastructure is to support and serve the end-user segment which is generally referred to as *Home Network*. This network includes any terminal devices which are located at the customer's site, such as home gateways. The power consumption in wired communication networks is listed in Table 2.1.

Figure 2.5 shows the wired Telecommunication access network, which contain three different physical layer technologies:

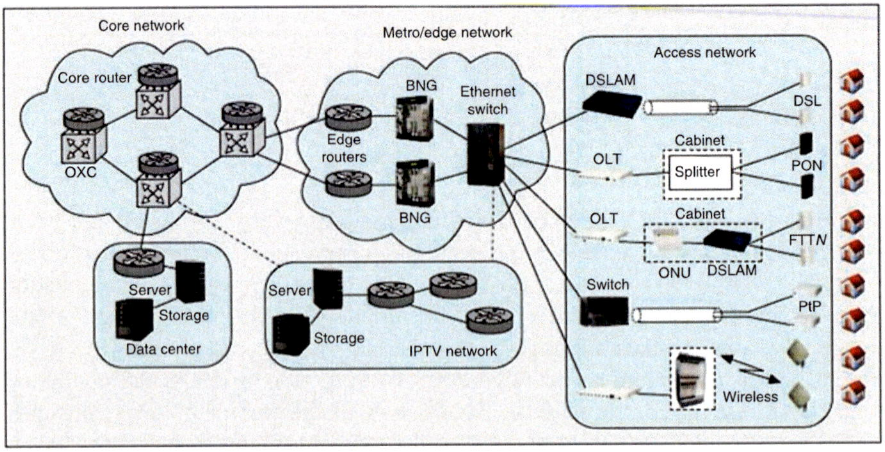

Fig. 2.5 Domains of wired communication networks [57]

- The first is considered to be the very widely deployed copper-based access technology, such as xDSL (Digital Subscriber Line).
- The second well-known technology is the fiber-based optical access network for broadband data transmission, generally named Fiber-To-The-x (FTTx).
- Another famous technology is the coaxial cable technology on which the DOCSIS (Data over Cable Service Interface Specification) standard is used.

Regarding the xDSL technologies, several types are used, which vary in bit rate and maximum range. The ADSL-type (Asymmetric DSL) and VDSL-type (Very high bit-rate DSL) flavors are the most common [47]. The most recent DSL technologies that are currently being deployed, namely ADSL2+ and VDSL2 can provide high data rates. In DSL technologies the last node before the subscriber is the DSLAM (DSL Access Multiplexer), it is situated in a remote node (RN) that is connected via an optical fiber to the Central Office (CO) equipment (FTTH), while DSL is then applicable on copper wires reaching from the remote node's DSLAM to the Customer Premises Equipment (CPE). The DSLAM acts like a network switch since its functionality is at Layer 2 of the OSI model [59]. On the other hand, there are different FTTx systems which depend on how close to the subscriber the fiber reaches [60]. A typical example is the Fiber-To-The-Home (FTTH), which means that the optical signal reaches the end user's home equipment. Other examples (as in Fig. 2.6) are Fiber-To-The-Building (FTTB), Fiber-To-The-Curb (FTTC), and Fiber-To-The-Node (FTTN) [61]. Moreover, FTTH networks can be classified as either AON (Active Optical Network) or PON (Passive Optical Network), depending on the use of passive or active devices. The most frequently used standard is GPON (Gigabit PON). Based on different network topologies, we can define FTTH networks as either point to point (P-t-P) or point to multipoint networks (P-t-MP).

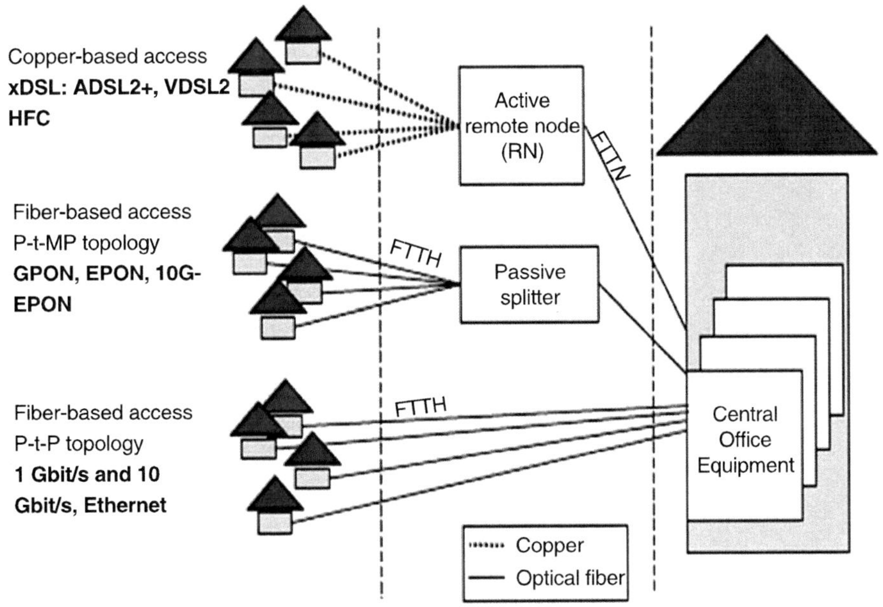

Fig. 2.6 Different wired access network technologies [62]

For passive optical networks, the connection of the customer's terminals is served by an OLT device (Optical Line Termination). A typical OLT can serve about 32 or 64 customers. P-t-P systems are implemented using direct connections between CPEs and COs usually through 1 Gbit/s or 10 Gbit/s Ethernet links [63].

2.3 Summary

This chapter analyzes and displays types of communication stations; the rate of consumption of electrical power by communication stations has also been addressed.

Chapter 3
Regenerative Fuel Cells as a Backup Power Supply

3.1 Introduction

Traditional telecom backup power solutions include batteries for short duration backup and diesel generators for more extended duration backup. Batteries are relatively inexpensive for 1–2 h of backup power. However, batteries are not ideal for more extended duration backup power applications because they can be expensive to maintain, unreliable after aging, temperature-sensitive and hazardous to the environment after disposal. Diesel generators are capable of more extended duration backup power. However, generators can be unpredictable, maintenance intensive, and emit high levels of pollution and greenhouse gases into the atmosphere. Fuel cells are reliable and quiet, with fewer moving parts than a generator, and a more extensive operating temperature range, −40 to +50 °C, than a battery. Also, a fuel cell system has a lower lifetime cost than a generator. The lower prices for the fuel cell are the result of only one maintenance visit per year and significantly higher system efficiency. Finally, the fuel cell is a clean technology solution with minimal environmental impact. Fuel cell systems provide backup power to critical communication network infrastructures in wireless, fixed, and broadband telecom applications ranging from 250 W to 15 KW, and they offer many outstanding features [64]. Fuel cells are currently finding use in various facets of technology. These devices are used in applications ranging from portable devices such as cell phones, where it replaces the battery supply, to automobile power trains and large-scale urban power stations. In each of these applications, the fuel cell is found to be a highly efficient, powerful and a remarkably clean form of power conversion. It is capable of transferring the energy stored in hydrogen or hydrogen-rich fuels into electrical energy with almost no emissions [65]. A fuel cell is an energy conversion device, which converts the chemical energy of a fuel and oxidant [66], often hydrogen and oxygen, to electrical energy. Fuel cells are similar to batteries; however, unlike battery a fuel cell must be continuously provided with fuel, rather than deriving energy from materials contained within the cell, and the products of the

A. A. Elbaset, S. Ata, *Hybrid Renewable Energy Systems for Remote Telecommunication Stations*, https://doi.org/10.1007/978-3-030-66344-5_3

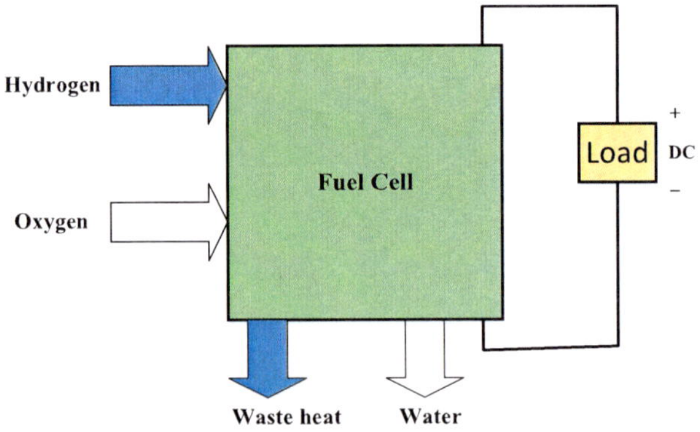

Fig. 3.1 A fuel cell's inputs and outputs

electrochemical reaction must be removed from the cell. It will produce DC electricity (plus water and heat) continuously [67], as shown in Fig. 3.1.

Benefits of Fuel Cells for Telecommunication [64]

1. **Autonomy:** Fuel cells can operate as long as there is available fuel, so whether an 8-h, 1-day or 3-day extended runtime is required, enough fuel can be stored on-site.
2. **Remote monitoring:** Fuel cells can be fully monitored from one central location alerting the operator as to when the system is in use and how long before refueling is required to ensure no downtime.
3. **Space requirement:** The space required for the same period of the runtime is considerably less for fuel cells than for battery banks. Fuel cells do not require cooling like batteries which eliminates the need for large cooling systems.
4. **Fuels:** The majority of these systems operate on hydrogen (in this instance the only emission is water), which can be generated from renewable sources (electrolysis) as shown in Fig. 3.2 or from reformed hydrocarbons (methanol, propane, ammonia, and natural gas).
5. **Temperature tolerance:** Unlike batteries, fuel cells do not degrade high temperatures and their range can be between −40 and +50 °C without any cooling required.
6. **Integration:** Fuel cell systems either provided as a stand-alone unit similar in size to a small refrigerator (for applications like base stations) or can be inserted in existing 19″ racks. So fuel cells are fit for outdoor as well as indoor applications.
7. **Cost:** Over the lifetime of the unit can offer cost savings over existing technologies. This includes maintenance, repairs, transport, and disposal.
8. **Reliability:** In many cases, fuel cells can offer higher reliability and MTBF (Mean Time Between Failures), and there is no degradation of voltage over time. Failures tend to be less critical and easily dealt with.
9. **Environmental:** Unlike generators, fuel cells do not use combustion, and therefore there are no NO_x, So_x, or particulate emissions from the unit. So the fuel cells provide clean energy and hence the green energy.
10. **Maintenance:** Fuel cells have very few moving parts which reduce the need for regular maintenance.

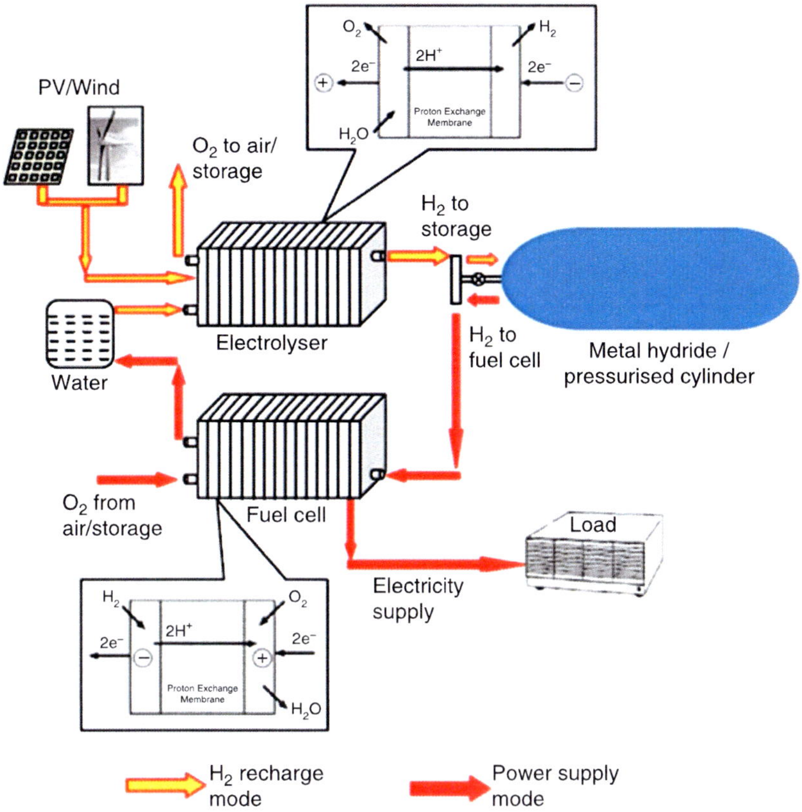

Fig. 3.2 DRFC system [82]

3.2 Types of Fuel Cells

The different fuel cell types are usually distinguished by the electrolyte that is used, though there are always other significant differences as follow [68]:

3.2.1 Polymer Electrolyte Membrane FC (PEMFC)

PEMFCs are suitable for either vehicles or power generation. They use a solid polymer membrane electrolyte and carbon electrodes, with platinum as a catalyst. PEMFCs are the best candidates for powering FCV as they operate at low temperature (80 °C) and offer short start-up time, high efficiency, and excellent power density. Operational efficiency is well below the theoretical value of 65%, but it is more than twice that of typical combustion engines, with high sensitivity to operating conditions [69]. The current PEMFC is less durable than combustion engines.

Membranes are sensitive to humidification. Anode catalysts are sensitive to poisoning by carbon monoxide and sulfur. They need rather pure hydrogen input from electrolysis or from reforming, with extensive clean-up. They also need cooling to avoid overheating. Current research efforts focus on high-temperature membranes and new catalysts to improve performance and reduce costs. PEMFC can also be used for distributed power generation, with a net efficiency of 35–40%. Potential breakthroughs in hydrogen distributed generation and synergies between PEMFC and PEM electrolyzer could make PEMFC a very attractive option [70].

3.2.2 Molten Carbonate FC (MCFC) and Solid Oxide FC (SOFC)

MCFC and SOFC are the best candidates for stationary power generation. MCFC are in most cases fueled by natural gas or biogas. They cannot be fueled by pure hydrogen as they need CO_2 input. SOFC can be powered by either hydrocarbons or hydrogen [71]. More information about this type can be found in [72].

3.2.3 Phosphoric Acid Electrolyte FC (PAFC) and Alkaline FC (AFC)

PAFC and AFC are also used for stationary applications. PAFC was the first FC ever commercialized. They tolerate H_2 impurities and offer 37–42% electrical efficiency (85% in co-generation). While many units are in operation, the potential market for PAFC is limited, since they are heavy, expensive ($4000–$4500/kW), and unlikely to become much cheaper. AFC was developed for power generation within the United States space program. They use the KOH solution as the electrolyte and non-precious metals as catalysts. They operate at 100–250 °C (recent versions at 23–70 °C) with an efficiency of 60%. The alkaline electrolyte has a low tolerance to CO_2, but AFC can also function in the air with filters [73].

3.2.4 Direct Methanol FC (DMFC)

DMFCs use methanol as a fuel. With low efficiency (15–30%) and low power density, they are not suitable for mobile or stationary use. Because methanol is easily transportable, however, they represent an option to replace batteries in portable devices [74]. In contrast, PEMFC offers limited benefits for use in portable devices, as H_2 storage offers no energy density advantage over batteries. Micro-SOFC is also being developed for portable use. Other FC concepts such as direct-ethanol FC (DEFC) are under development for use in FCV [75].

3.2.5 Alkaline Fuel Cell (AFC)

AFCs are long used by NASA on space missions; alkaline fuel cells can achieve power generating efficiencies of up to 70%. They were used on the Apollo spacecraft to provide both electricity and drinking water. Alkaline fuel cells use potassium hydroxide as the electrolyte and operate at 160 °F. However, they are very susceptible to carbon contamination, so they require pure hydrogen and oxygen [76].

3.2.6 Zinc-Air Fuel Cell (ZAFC)

In a typical zinc/air fuel cell, there is a gas diffusion electrode (GDE), a zinc anode separated by electrolyte, and some form of mechanical separators. The GDE is a permeable membrane that allows atmospheric oxygen to pass through. After the oxygen has converted into hydroxyl ions and water, the hydroxyl ions will travel through an electrolyte and reaches the zinc anode. Here, it reacts with the zinc and forms of zinc oxide. This process creates an electrical potential; when a set of ZAFC cells are connected, the combined electrical potential of these cells can be used as a source of electric power [77].

3.2.7 Protonic Ceramic Fuel Cell (PCFC)

This new type of fuel cell is based on a ceramic electrolyte material that exhibits high protonic conductivity at elevated temperatures. PCFCs share the thermal and kinetic advantages of high-temperature operation at 700 °C with molten carbonate and solid oxide fuel cells, while exhibiting all of the intrinsic benefits of proton conduction in PEM and phosphoric acid fuel cells. The high operating temperature is necessary to achieve very high electrical fuel efficiency with hydrocarbon fuels [78].

3.2.8 Microbial Fuel Cell (MFC)

Microbial fuel cells use the catalytic reaction of microorganisms such as bacteria to convert virtually any organic material into fuel. Some common compounds include glucose, acetate, and wastewater. Enclosed in oxygen-free anodes, the organic compounds are consumed (oxidized) by the bacteria or other microbes [73]. As part of the digestive process, electrons are pulled from the compound and conducted into a circuit with the help of an inorganic mediator. MFCs operate well in mild conditions relative to other types of fuel cells, such as 20–40 °C, and could be capable of producing over 50% efficiency. These cells are suitable for small-scale applications such as potential medical devices fueled by glucose in the blood, or larger such as water treatment plants or breweries producing organic waste that could then be used to fuel the MFCs [79].

3.2.9 Regenerative Fuel Cells

The fuel cells presented above convert energy one way: from fuel to electricity. They are not designed for the reverse recharging operation. Recharging the fuel cell requires an electrolyzer (EZ) to decompose the water back into hydrogen and oxygen. An electrolyzer is generally a separate unit from the fuel stack, and the two cannot operate simultaneously. The coupling of FC stack and an EZ stack is conventionally known as Regenerative Fuel Cell (RFC) [80].

3.3 Principles of Regenerative Fuel Cells

A Regenerative Fuel Cell (RFC) is a single device or system capable of functioning either as an electrolyzer or a fuel cell. The reversible reaction most commonly used in RFC is the splitting of water into its constituent elements, hydrogen and oxygen, and their recombination to form water once again [81]:

$$H_2O + \text{energy} \leftrightarrow H_2 + \tfrac{1}{2}O_2 \tag{3.1}$$

Other reactions that have been investigated for RFCs are hydrogen-halogen reactions (in particular bromine) and the zinc-oxygen reaction. It will be focusing on RFCs based on the water-splitting reaction in this book. In RFC based on this reaction, electrical input to the cell in electrolyzer (EZ) mode is used to dissociate liquid water into hydrogen and oxygen gas. These gases are stored and fed back into the RFC acting now in fuel cell (FC) mode to regenerate electricity and reform water. Alternatively, just the hydrogen is stored, and the RFC draws on oxygen from the air when operating as a fuel cell [82]. The RFC concept is thus founded upon the inherent reversibility of the water dissociation and reformation (or similar) reaction.

Coupled to hydrogen and oxygen storage systems, or just a hydrogen storage system if the air is used to supply oxygen, RFC constitutes an integrated system for storing electrical energy for later use [81]. RFCs, therefore, have potential applications in areas such as space vehicles and satellites, submarines, aircraft, remote or distributed terrestrial energy supply systems, electricity storage in central grids or at a local level when there is variable energy input from solar or wind power electricity storage at a local level and possibly in hydrogen-electric cars capable of generating some of their hydrogen fuel when parked [83]. RFCs come in two basic types:

1. Discrete Regenerative Fuel Cells (DRFC).
2. Unitized Regenerative Fuel Cells (URFC).

3.3.1 Discrete Regenerative Fuel Cells (DRFC)

In a discrete regenerative fuel cell (DRFC), the device used to perform the electrolysis is separate from that used as a fuel cell, but these two devices are both integrated into a single system as in Fig. 3.2. The overall DRFC also includes a water management system, a control system, and a gas storage system if it is to function as a stand-alone energy storage unit [82]. One key advantage of a DRFC over a separate electrolyzer, hydrogen storage, and fuel cell is that by integrating all functional units into a single compact system the DRFC can provide a stand-alone alternative technology for storing electrical energy, and hence compete directly with batteries. Additionally, a DRFC allows the closing of the water cycle, whereby product water in the FC part can be returned to the water storage for the electrolyzer [82–84]. This capability is particularly important for space, aircraft, or submarine applications where water must be conserved by recycling. It may become increasingly beneficial in stationary power supply applications too in areas where supplies of fresh water are very limited. Keeping the electrolyzer and fuel cell separate allows these two components to be designed to optimize their individual performance and cost-effectiveness. It also allows the capacities of the cell in the two modes to be set independently, and the cell to operate in both modes simultaneously (if the overall system designs permit) [81]. The overall design of the DRFC system consists of a water management system, a control system, and a gas storage system to function as a stand-alone energy storage unit [82]. The main advantage of a DRFC over a separate conventional electrolyzer, hydrogen storage.

3.3.2 Unitized Regenerative Fuel Cells

The Unitized Regenerative Fuel Cell (URFC), also known as reversible PEM fuel cell, or reversible regenerative fuel cells, as displayed in Fig. 3.3, refine this concept by using the same cell electrodes to perform both the electrolyzer function (equivalent to battery charging) and the fuel cell function (equivalent to battery discharging). A Unitized Regenerative Fuel Cell System (URFCS) incorporates the URFC into an overall energy storage system [85, 86].

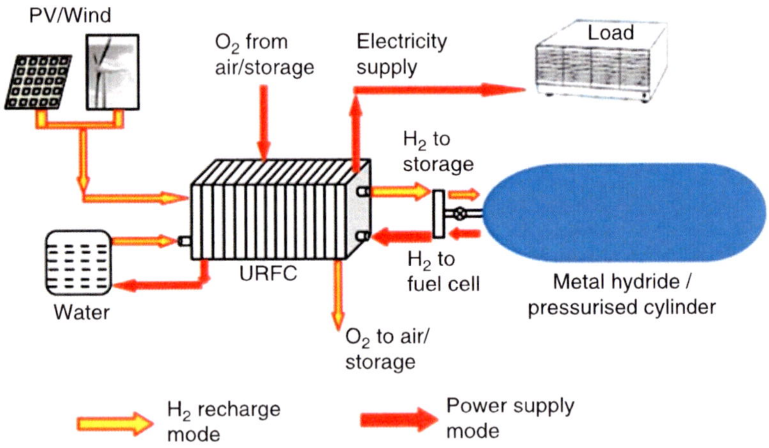

Fig. 3.3 URFC system [82]

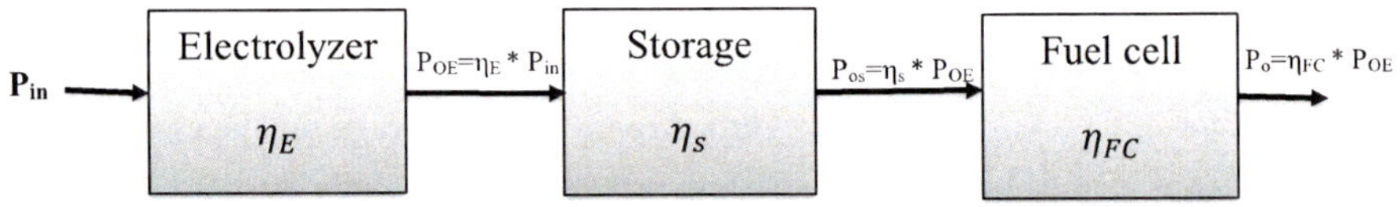

Fig. 3.4 Scheme diagram of powers and efficiencies in RFC

3.4 Efficiency of Regenerative Fuel Cell

If RFC is used in a system to store electrical energy in the form of hydrogen and subsequently regenerate electricity from the hydrogen stored, the roundtrip energy efficiency of the overall energy storage system is displayed in Fig. 3.4.

From Fig. 3.4 P_{in} is the input power to RFC, η_{E} is the electrolyzer efficiency, P_{OE} is the out power contained in hydrogen from electrolyzer, η_{s} is the storage efficiency, P_{os} is the output power from storage, η_{FC} is the fuel cell efficiency, and P_{o} is the output and use electric power from RFC, where η_{s} measures the net energy efficiency of the hydrogen storage system, taking into account any energy required to get the hydrogen into and out of the storage (for example, to compress the hydrogen in the case of high-pressure hydrogen gas storage, and friction losses in the form of heat on exit), and any losses of hydrogen from the storage over time. Since η_{s} will always be less than 100%,

$$P_{\text{OE}} = \eta_{\text{E}} * P_{\text{in}} \tag{3.2}$$

$$P_{\text{os}} = \eta_{\text{s}} * P_{\text{OE}} \tag{3.3}$$

$$P_{\text{o}} = \eta_{\text{FC}} * P_{\text{OE}} \tag{3.4}$$

Recombine the last three equations in one will be given:

$$P_o = (\eta_{E*}\ \eta_{s*}\ \eta_{FC}) * P_{in} \tag{3.5}$$

Thus

$$\eta_{RFC} = \eta_E * \eta_{FC} * \eta_s \tag{3.6}$$

URFC has the same cell as employed in both modes, and hence the ratio of EZ-mode to FC-mode capacity is fixed by the cell characteristics. Generally, it is difficult to optimize the efficiency of this single cell in both EZ and FC modes, because different catalysts are required on the oxygen side for optimum performance and a mixed catalyst has, therefore, to be used [81].

The roundtrip energy efficiency of a URFC is therefore usually less than that of a comparable DRFC. Moreover, since the same cell is employed in both modes, a URFC can only operate in one mode at a time, either EZ or FC mode. Both the DRFC and URFC systems are limited by the unavoidable energy losses involved in converting electrical energy to hydrogen gas (that is stored, chemical energy) by splitting water, and then converting the chemical energy of the hydrogen gas back into electrical power by reacting it with oxygen in a fuel cell [81].

3.5 Methodology and Emphasis

It is required to design the fuel cell power system for supplying communication stations with electric energy as a backup power system. This book focuses on DRFCs, principles of RFCs are discussed, DRFC is suitable for remote communication station to the shortage of water supply, higher efficiency related to URFC, EZ part, and FC can be worked at the same time.

The DRFC power system is used in this study. In this system, there is an electrolyzer part and fuel cell part. It is required to determine the input energy to Electrolyzer ($P_{in_{el}}$) [66], amounts (quantities) of hydrogen and oxygen production, and size of the hydrogen tank.

3.5.1 Hydrogen Production Estimation

Hydrogen production must be sufficient for feeding fuel cells electric energy generated by FC feeds loads at deficit times. The hydrogen production from electrolyzer equals fuel cells hydrogen consumed and of course, the generated electrical power from fuel cell must be equal the deficit.

$$P_{in_{el}}(t) = \frac{P_{sur}(t)}{(\eta_E * \eta_s * \eta_{FC})} \tag{3.7}$$

Then, with knowing $P_{sur}(t)$, η_E, η_s, and η_{FC} from specs of electrolyzer part, the hydrogen production can be calculated as follows:

For each mole of hydrogen production

$$\text{Charge} = 2F * \text{amount of } H_2 \tag{3.8}$$

Dividing by time and rearranging

$$H_2 \text{ prod.} = \frac{I}{2F} \tag{3.9}$$

This is for a single cell, for a stack of n cells

$$H_2 \text{ prod.} = \frac{In}{2F} \tag{3.10}$$

However, it would be more useful to have the formula in kg/s, without needing to know the number of cells, and in terms of power [66], rather than current. If the voltage of each cell in the stack is V_c, then

$$\text{Power}, P_{in_{el}} = V_c \times I \times n \tag{3.11}$$

So

$$I = \frac{P_{in_{el}}}{V_c \times n} \tag{3.12}$$

Substituting equations (3.11, 3.12) into equation (3.10) gives

$$H_2 \text{ prod} = \frac{P_{in_{el}}}{2 \times V_c \times F} (\text{mole/s}) \tag{3.13}$$

Changing from mole/s to kg/s, the molar mass of hydrogen is 2.02×10^{-3} kg/mole, so this becomes

$$H_2 \text{ prod} = \frac{2.02 \times 10^{-3} \times P_{in_{el}}}{2 \times V_c \times F} \tag{3.14}$$

$$= 1.05 \times 10^{-8} \times \frac{P_e}{V_c} (\text{kg/s}) \tag{3.15}$$

For converting hydrogen production to m^3/h divided by hydrogen density ($\rho H_2 = 0.084$ kg/m^3) and multiply by 3600 to have hydrogen production per hour

$$H_2 \text{ prod} = \frac{1.05 \times 10^{-8} \times P_{in_{el}} \times 3600}{V_c \times \rho_{H_2}} \tag{3.16}$$

$$= \frac{1.05 \times 10^{-8} \times P_{\mathrm{in_{el}}} \times 3600}{V_{\mathrm{c}} \times 0.084} \tag{3.17}$$

$$= 0.00045 \frac{P_{\mathrm{in_{el}}}}{V_{\mathrm{c}}} \left(\mathrm{m^3/h}\right) \tag{3.18}$$

The value of V_c can be calculated as:

$$V_{\mathrm{c}} = \frac{\mu_{\mathrm{E}} * 1.48}{\eta_{\mathrm{E}}} \tag{3.19}$$

then

$$\mathrm{H_2\ \ prod} = \frac{0.00045 \times P_{\mathrm{in_{el}}} \times \eta_{\mathrm{E}}}{\mu_{\mathrm{E}} \times 1.48} \tag{3.20}$$

$$= 3.04054 \times 10^{-4} \times \frac{\eta_{\mathrm{E}} \times P_{\mathrm{e}}}{\mu_{\mathrm{E}}} \left(\mathrm{m^3/h}\right) \tag{3.21}$$

A reasonable estimate for μ_f is 0.95.

3.5.2 *Oxygen Production Estimation*

Oxygen production can be determined as the above method, so:

$$\mathrm{O_2\ \ prod} = 1.52027 \times 10^{-4} \times \frac{\eta_{\mathrm{E}} \times P_{\mathrm{e}}}{\mu_{\mathrm{E}}} \left(\mathrm{m^3/h}\right) \tag{3.22}$$

The water consumption is 1 l/N m^3 of hydrogen production.

3.5.3 *Hydrogen Usage in the Fuel Cell Part*

From the last analysis and the basic operation of the fuel cell, hydrogen usage will be:

$$\mathrm{H_2\ \ usage} = 0.00045 \times \frac{P_{\mathrm{eo}}}{V_{\mathrm{c}}} \left(\mathrm{m^3/h}\right) \tag{3.23}$$

H_2 usage can be obtained by substituting V_c from Eq. (3.19) in Eq. (3.23), then *H_2 usage* will be as follows:

$$\text{H}_2\ \text{usage} = \frac{0.00045 \times P_{\text{eo}} \times \mu_{\text{f}}}{\eta_{\text{FC}} \times 1.48}\left(\text{m}^3/\text{h}\right) \tag{3.24}$$

$$= 3.04054054 \times 10^{-4} \times \frac{P_{\text{eo}} \times \mu_{\text{f}}}{\eta_{\text{FC}}}\left(\text{m}^3/\text{h}\right) \tag{3.25}$$

Oxygen usage can be determined as above, so:

$$\text{O}_2\ \text{usage} = 1.52027027 \times 10^{-4} \times \frac{P_{\text{eo}} \times \mu_{\text{f}}}{\eta_{\text{FC}}}\left(\text{m}^3/\text{h}\right) \tag{3.26}$$

3.5.4 *Air Usage in Fuel Cell Part*

Oxygen about 21% of air, so:

$$\text{Air usage} = 1.52027027 \times 10^{-4} \times \frac{P_{\text{eo}} \times \mu_{\text{f}}}{0.21 \times \eta_{\text{FC}}}\left(\text{m}^3/\text{h}\right) \tag{3.27}$$

$$\text{Air usage} = 7.2393822 \times 10^{-4} \times \frac{P_{\text{eo}} \times \mu_{\text{f}}}{\eta_{\text{FC}}}\left(\text{m}^3/\text{h}\right) \tag{3.28}$$

However, if the air were used at this rate, then as it left the cell it would be completely devoid of any oxygen—it would all have been used. This is impractical, and in practice, the airflow is well above stoichiometry, typically twice as much. If the stoichiometry is λ, then the equation for air usage becomes

$$\text{Air usage} = 7.2393822 \times 10^{-4} \times \lambda \times \frac{P_{\text{eo}} \times \mu_{\text{f}}}{\eta_{\text{FC}}}\left(\text{m}^3/\text{h}\right) \tag{3.29}$$

3.5.5 *Air Exit Flow Rate*

It is sometimes essential to distinguish between the *inlet* flow rate of the air and the *outlet* flow rate, so:

$$\text{Exit air flow rate} = \text{Air inlet flow rate} - \text{oxygen usage} \tag{3.30}$$

$$\text{Exit air flow rate} = (7.2393822 \times \lambda - 1.52027027) \times 10^{-4} \times \frac{P_{\text{eo}} \times \mu_{\text{f}}}{\eta_{\text{FC}}} \times \left(\text{m}^3/\text{h}\right) \tag{3.31}$$

3.6 Hydrogen Storage

In general, there are three methods for storing hydrogen, i.e., (1) gaseous hydrogen storage, (2) liquid hydrogen storage, and (3) metal hydride storage [87]. In the renewable hydrogen power system, the simplest and most practical way to store the hydrogen gas is by compressing it, so a cylindrical storage tank is used for the compressed hydrogen. The main parameters in designing the hydrogen gas storage were volume, pressure, and temperature. A cylindrical hydrogen storage tank was chosen. Such a reservoir could be situated underground for safety reasons and to avoid excessive temperature fluctuations [88].

Since hydrogen behaves very much like an ideal gas in the ambient temperature. Therefore, the mathematical model for the hydrogen pressure p in a storage tank can be calculated from ideal gas law as follow [89]:

$$P = \frac{nRT}{V} \tag{3.32}$$

where P *is* hydrogen pressure inside the tank, P_a, n is the number of moles, mol. R is the universal gas constant (8.31451), J/K/mol. T *is the* temperature of the gas, K. V is the volume of the tank, m^3.

Assume constant temperature then:

$$P_i \times V_i = P_f \times V_f \tag{3.33}$$

where P_i is the initial pressure (bar). P_f *is* the final pressure (bar). V_i is the initial volume (m^3). V_i is the final volume (m^3).

Then

$$V_f = \frac{P_i \times V_i}{P_f} \tag{3.34}$$

More information about hydrogen storage can be found in [90–93].

3.7 Materials Balance in DRFC

Hydrogen generated from electrolyzer mode must be sufficient for storage and feed RFC at fuel cell mode to generate electric energy that can be fed load during defect hours.

3.7.1 Hydrogen Balance

Hydrogen is one of the products of electrolyzer (mode or part). Hydrogen is stored in special tanks for reuse to generate electricity at deficit hours, so the amount of generated hydrogen must be sufficient for generating electric energy that required in deficit hours, so hydrogen produced from the electrolyzer part is equal to hydrogen consumed by fuel cell during the year hours.

$$\sum_{i=1}^{i=8760} \left(H_{2\text{consum.}} - H_{2\text{prod.}}\right) = 0 \tag{3.35}$$

3.7.2 Water Balance

There is a water closed cycle. Water is used by EZ (mode or part) in RFC to generate hydrogen and oxygen; then hydrogen is stored in special tanks, also in a certain application, oxygen is stored in other containers for reuse. Hydrogen and oxygen are used to generate electricity from fuel cell (mode or part) during year hours:

$$\sum_{i=1}^{i=8760} \left(\text{water}_{\text{consum.}} - \text{water}_{\text{prod.}}\right) = 0 \tag{3.36}$$

Also, consumed water equals product water during the year.

Water tank storage size equals the maximum accumulated water during the year.

3.7.3 Oxygen Balance

Oxygen is one of the products of electrolyzer (mode or part). Oxygen is stored in tanks in specific applications for reuse in a fuel cell (mode or part) to generate electricity at deficit hours, so amount of generated oxygen must be sufficient for generating electric energy that required in deficit hours, so oxygen produced from electrolyzer part is equal to oxygen consumed by fuel cell during the year hours.

$$\sum_{i=1}^{i=8760} \left(O_{2\text{consum.}} - O_{2\text{prod.}}\right) = 0 \tag{3.37}$$

3.8 Summary

This chapter presents the principles of fuel cells, the idea of their functioning and their internal components—the benefits of fuel use in communications.

Regenerative fuel cells (RFCs) have been selected in this study. Hydrogen and oxygen output and their relationships to the energy consumed are explained. The rates of used hydrogen for electricity production were also explained. Calculations of the water and hydrogen tanks and ways of hydrogen storage are displayed. How to get RFC efficiency was explained.

Chapter 4
Optimum Sizing of Ultracapacitors

4.1 Introduction

Renewable energies require efficient and reliable energy storage. Although renewable energy is free and environment-friendly source of electricity, a storage element is needed as an energy buffer in wind and photovoltaic systems to bridge the gap between available and required energy [94], also, for uninterrupted power supply (UPS) systems when the storage device can be used for temporary backup power in UPS systems. They can provide instantaneous supplies of electricity without delays, helping to prevent malfunctions of mission-critical applications [95]. The batteries are generally the most popular energy storage device, because of its low cost and wide availability. However, photovoltaic panels or wind are not an ideal source for battery charging as the output is unreliable and heavily dependent on weather conditions.

The batteries cannot get an immediate response to the load requirement when sudden clouds or no wind occur. The batteries are often deep discharged, which damages the battery and shortens its useful life. It is not possible to ensure an optimum charge/discharge cycle. Undercharging of the battery leads to sulfating and stratification, both of which shorten the lifetime of the battery. Another cause of reduced battery life is gassing, which results from battery overcharging [94]. For avoiding all previous problems new storage devices should be used, one of these devices is ultracapacitor (UC). This device is a kind of capacitor.

The capacitors are fundamental electric components characterized by its ability to store energy in an electric field developed through the accumulation of electric charge [96]. The capacitor's ability to accumulate electric charge and store electric energy is defined by its capacitance.

Capacitors can, in general, be divided into three general categories: electrostatic, electrolytic, and electrochemical. The electrostatic capacitor is the conventional capacitor, consisting of two conducting plates with an isolating dielectric between the plates. An electrolytic capacitor employs a conductive electrolytic salt in direct

A. A. Elbaset, S. Ata, *Hybrid Renewable Energy Systems for Remote Telecommunication Stations*, https://doi.org/10.1007/978-3-030-66344-5_4

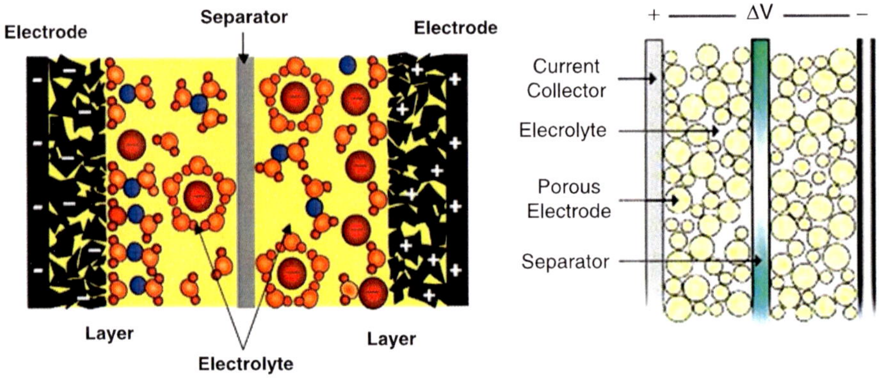

Fig. 4.1 Internal composition of ultracapcitor cell

contact with the electrodes, instead of a dielectric. This reduces the effective plate separation and thereby increases the capacitance of the capacitor. As an advanced version of the former, as in Fig. 4.1, the electrochemical capacitor employs sophisticated porous electrodes with an electrolyte in between, which increases the capacitance even more [97]. Ultracapacitors can be an excellent replacement for common batteries where applications request power burst, quick charging, temperature stability, and excellent safety properties (immunity to shock and vibration). Electrochemical capacitors have significant advantages for deployment in renewable energy resource systems, like the photovoltaic systems. Some of these are listed below according to [95]:

Lack of maintenance: In contrast to the battery maintenance, capacitors require no maintenance. This greatly reduces system cost over time and allows the storage system to be located in places impractical for chemical battery systems (e.g., buried).

Longevity: Because capacitors store charge physically rather than chemically, cycling has virtually no effect on their capacity or longevity. A 20-year life is easily achieved by the proper selection of materials and control of operating parameters.

Environmentally benign: Capacitors do not employ toxic materials and thus present no environmental threat in the manufacture, transport, or disposal. They do not outgas in use and present no threat of explosion.

High discharge rate capability: Capacitors can be discharged at very high rates without damage. High rates, however, reduce the delivered energy of the unit.

A comparison of properties of the rechargeable lead-acid battery, conventional capacitor, and electrochemical capacitor are presented in Table 4.1 [98].

Ultracapacitors have several practical applications, including [99]:

Transportation: The most promising market for UC is in the transportation industry. They can be used in automobiles by coupling them with other energy sources,

Table 4.1 Comparison between the lead-acid battery, ultracapacitor, and conventional capacitor

Available performance	Lead acid battery	Ultracapacitor	Conventional capacitor
Charge time	1–5 h	0.3–30 s	10^{-6} to $10^{-}3$ s
Discharge time	0.3–3 h	0.3–30 s	10^{-6} to 10^{-3} s
Energy (Wh/kg)	10–100	1–10	<0.1
Cycle life	1000	>500,000	>500,000
Specific power (W/kg)	<1000	<10,000	<100,000
Charge/discharge efficiency	0.7–0.85	0.85–0.95	>0.95
Operating temperature (°C)	−20 to 100	−40 to 65	−20 to 65

particularly batteries. They can also improve fuel efficiency by storing energy when an automobile is braking and then releasing the energy when it accelerates. Automobiles powered by coupling fuel cells with UC are ideal choices for stop-and-go traffic where UC provides sudden bursts of energy during start-up and fuel cells provide sustained energy. UCs also perform well at extremely low temperatures (–40 °C), are small and have lightweight designs.

Consumer electronics: UCs are widely used in consumer electronics as backup energy sources for system memories, microcomputers, system boards and clocks, toys, and mobile phones. They are ideal for devices requiring quick charges. UCs are cost-effective options because they have extremely long lifetimes and do not need replacement during the lifetimes of the devices they power.

Uninterrupted power supply (UPS) systems: UCs quickly bridge the power applications for stationary UPS systems that are augmented with fuel cells. In addition, UCs are best suited to provide power for start-up and during peak load buffering.

Other applications: UCs show promise for critical-load operations such as hospitals, banking centers, airport control towers, and cell phone towers. The critical time between a power outage and the start of a generator can be bridged effectively by UCs because they provide power within milliseconds to a few seconds after an interruption.

In this chapter, UC characteristics, UC modeling, and sizing of UC are introduced.

4.2 Ultracapacitor Characteristics

The ultracapacitor characteristics are determined by temperature and voltage dependencies, frequency, self-discharge and charge redistribution. In this study temperature and voltage, dependencies are taken into account for estimating UC sizing.

4.2.1 Temperature Dependency

Many applications that utilize UC, for example, hybrid vehicles, operate at different environmental temperatures. It is mainly the ionic resistance part of the ESR that is influenced by temperature due to changes in the viscosity at low temperatures below 0 °C. The capacitance, on the other hand, shows little dependence on the temperature at optimal frequency operation but decreases some in the middle frequency range [100]. ESR and capacitance dependence on temperature can be seen in Fig. 4.2.

A temperature correction factor (TCF) can be used to define the effects of temperature changes on ESR. The equation of TCF deducted from the resistance relative performance curve can be defined as [101]:

$$\text{TCF} = -3.5 * 10^{-7}\, T^3 + 7.7 * 10^{-5}\, T^2 - 0.0054 * T^3 + 1.1 \tag{4.1}$$

where T is the temperature of UC under operation.

Then R as a function of temperature can be defined as:

$$R(T) = \text{TCF} * \text{ESR} \tag{4.2}$$

4.2.2 Voltage Dependency

Contrary to a conventional capacitor, the capacitance of UC varies with the terminal voltage. A typical voltage-dependent capacitance profile can be seen in Fig. 4.3. This could be explained by a reduction of the distance separating the charges at the phase

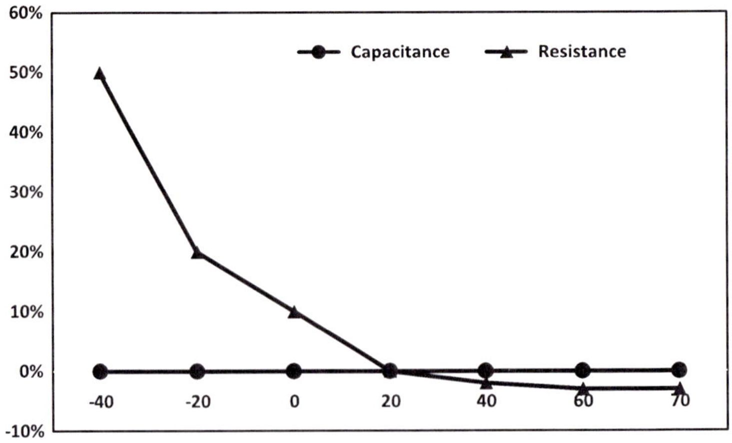

Fig. 4.2 ESR and capacitance percentage variation due to temperature

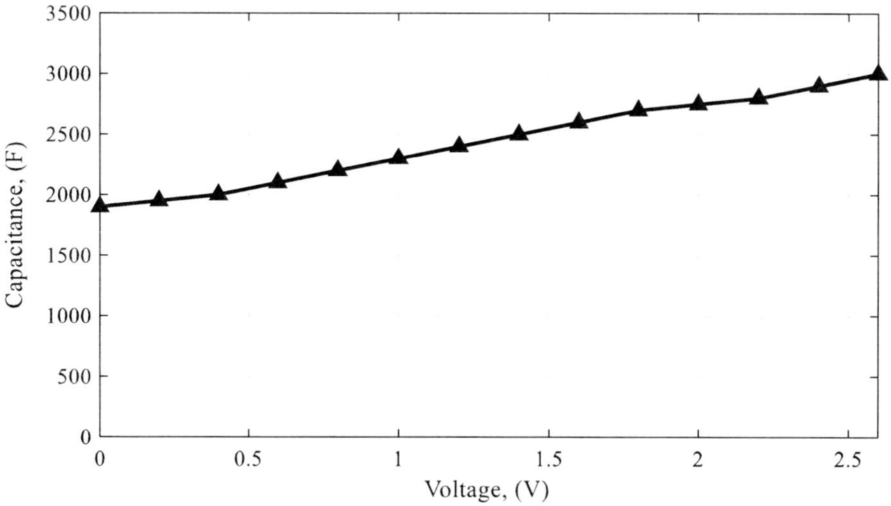

Fig. 4.3 Capacitance as a function of voltage for a typical UC

boundary between electrode and electrolyte as the voltage increases. In the formulation of adequate equivalent circuits, one should, therefore, take into account the variable capacitance [102].

The voltage-dependent capacitance is expressed through the so-called differential capacitance C_{diff} defined as the change in charge due to a change in voltage [103]:

$$C(v)_{\mathrm{diff}} = \frac{\mathrm{d}Q}{\mathrm{d}V} \tag{4.3}$$

Dividing with the incremental time interval (dt) results in the following relationship:

$$\frac{\mathrm{d}V(t)}{\mathrm{d}t} = \frac{1}{C(v)_{\mathrm{diff}}}.i \tag{4.4}$$

One can utilize several different methods to derive $C(V)_{\mathrm{diff}}$, where the most used method is a linear approximation resulting in an expression on the form:

$$C(V) = C_0 + K_{\mathrm{v}}V \tag{4.5}$$

where V is the applied voltage, C_0 is the capacitance close to zero voltage, and K_{v} is a constant slope (F/V) [104].

Ultracapacitor Stacking

In some practical applications, UC systems with both high energy and power are necessary. To generate more energy and power, several single ultracapacitors can be stacked together, and several such stacks can be connected to form an energy bank. The two options for stacking are in series and in parallel [105]. The behavior of connected UC in series and parallel is the conventional capacitor.

4.3 Ultracapacitor Electrical Models

In the application, an equivalent model which reflects characteristics of UC based power system can be described in the following items.

4.3.1 RC Serial Model

The simplest ultracapacitor circuit model is the RC model, which has only one RC branch. Figure 4.4 shows UC as a simple RC model. This model is composed of a resistor *R*, which models the ultracapacitor's ohmic loss, usually called equivalent series resistor (ESR) and a capacitor *C*, which simulates the ultracapacitor's capacitance during charging and discharging effects [106, 107].

Figure 4.5 shows the simulation results of the simple RC model. The charging and discharging current are constant at 30 A. The capacitor rated voltage, capacitance, and ESR are 2.5 V, 470 F, and 2.5 mΩ, respectively [108].

By comparing the simulation result and the experimental results, it can be seen that the simple RC model has several advantages and disadvantages. The primary

Fig. 4.4 Simple UC model

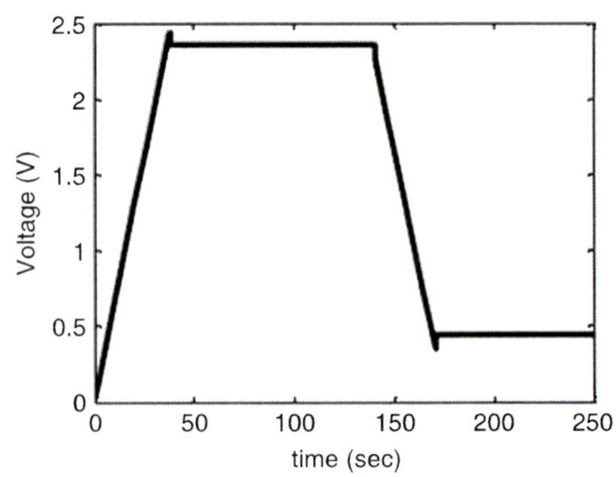

Fig. 4.5 Experimental test results of charge and discharge UC at constant current (30A)

advantage is the simplicity of the model. It is easy to incorporate into a circuit and the simulation process is computationally straightforward.

The primary disadvantage is that the simple RC model is not able to capture the nonlinear rise and fall of the ultracapacitor voltage and the change in voltage after the charging and discharging stops. Therefore, more detailed models are required for better accuracy.

4.3.2 First-Order Model for UC

Figure 4.6 shows the schematic circuit diagram as a representation of the first-order model for UC given by [94].

There are four ideal elements of the circuit given in Fig. 4.6, namely a capacitance C, a series resistor R_s, a parallel resistor R_p, and a series inductor L. During charging and discharging, series resistance which is also called the equivalent series resistance (ESR) contributes to energy loss. Leakage current resistance which is the parallel resistance R_p also takes energy loss due to capacitor self-discharge. In a practical capacitor, R_p is always much higher than R_s, that is why R_p can be neglected particularly in high-power applications.

There are many other UC models [94, 106] which dedicated according to the application of UC.

4.4 Methodology of UC Sizing

The bridge power system between the main supply and standby supply is the main function of UC. In this study, the results of sizing UC are a certain stack of UC cells (a type of UC cell, no. of series, and no. of parallel cells). There are two kinds of sizing: the first is a simple sizing that depends on the simplest equivalent circuit of UC and does not take into account the effect of temperature and voltage variation. The second is an accurate sizing that depends on taking into account the effect of temperature and voltage variation.

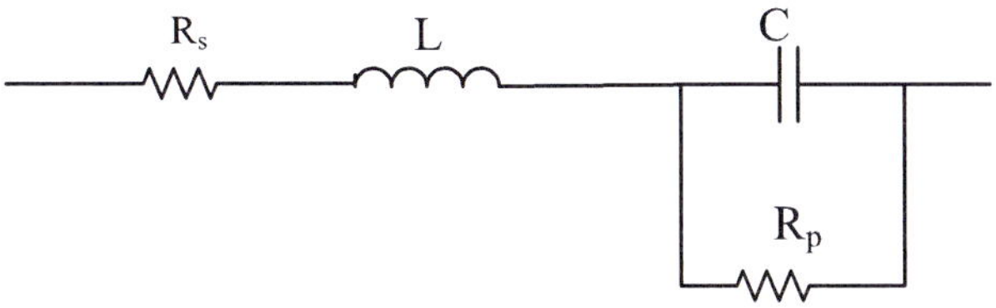

Fig. 4.6 The first-order circuit model of UC

4.4.1 Simple Sizing of UC

In order to size the appropriate UC cell for any application, we will need to determine the system variables required. Using this information we can calculate the proper size and number of cells required. In order to get a complete solution, the following parameters will need to be defined [109]:

- V_{max}: Maximum charged voltage
- V_w: Working voltage
- V_{min}: Minimum voltage
- P: Required power or I: application rated current
- t_d: Duration of discharge

Most applications can be categorized into two categories; the first one is constant current applications, and the second one is constant power applications, Fig. 4.7 displays the discharge curve of UC cell. This curve displays what happens in the time of supplying the load.

In the case of load, power remains constant so that the ultracapacitor discharge and the load voltage drops, the current increases to maintain load power (V_{Load} * I_{Load}) as constant [109]. Power P (assuming constant power) so:

$$P = V_{max} \cdot I_{min} = V_{min} \cdot I_{max} \tag{4.6}$$

and

$$I_{max} = \frac{P}{V_{min}} \tag{4.7}$$

$$I_{min} = \frac{P}{V_{max}} \tag{4.8}$$

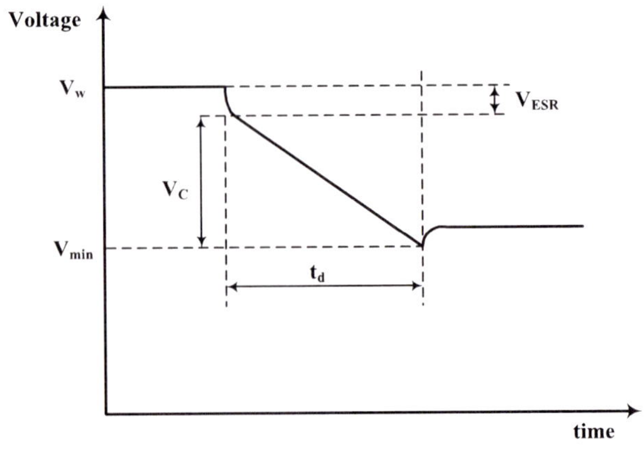

Fig. 4.7 Discharge curve of UC cell

$$I = \frac{I_{\max} + I_{\min}}{2} \tag{4.9}$$

where $I_{\max}$ is the maximum load current, $I_{\min}$ is the minimum load current, and I is the average load current.

During a discharge cycle, the initial drop in voltage is due to the Equivalent Series Resistance of the part (ESR). The amount of drop is a function of the ESR and discharge current as indicated by the equation below:

$$V_{\text{ESR}} = I * \text{ESR} \tag{4.10}$$

After the initial instantaneous drop due to ESR, the capacitor will discharge according to its capacitance and discharge current, as indicated in Eq. (4.5):

$$V_{\text{c}} = I * \frac{t_{\text{d}}}{C} \tag{4.11}$$

By placing these two equations together, the total voltage drop can be calculated per Eq. (4.6):

$$\text{d}V_{\text{total}} = V_{\text{c}} + V_{\text{ESR}} = \left(I * \frac{t_{\text{d}}}{C}\right) + (I * \text{ESR}) \tag{4.12}$$

where

$\text{d}V_{\text{Total}}$ = the drop in voltage when the capacitor is discharged. This is the difference between the V_{w} and $V_{\min}$ as indicated in Fig. 4.15.
I = Current in Amps used to discharge the capacitors.
t_{d} = Duration in Seconds to discharge the capacitor between V_{w} and $V_{\min}$.
C = the total capacitance of the ultracapacitor.

Values of I, t_{d}, V_{w}, and $V_{\min}$ are known from the application so C is calculated. If a single cell is used, then it is the cell capacitance. Ultracapacitor cell is typically rated at 2.7 V [109] so a number of series capacitor can be calculated as:

$$N_{\text{s}} = \frac{V_{\text{w}}}{V_{\text{c}}} \tag{4.13}$$

where N_{s} is the number of series cells and V_{c} is the cell rated voltage.

$$C_{\text{cell}} = C * N_{\text{s}} \tag{4.14}$$

If the rated current of each cell is smaller than load rated current, one or more parallel paths must be connected, its number is

$$N_{\mathrm{p}} = \frac{I}{I_{\mathrm{c}}} \tag{4.15}$$

So the total number of cells is:

$$C_{\mathrm{t}} = C_{\mathrm{cell}} * \frac{N_{\mathrm{p}}}{N_{\mathrm{s}}} \tag{4.16}$$

And total equivalent ESR is

$$\mathrm{ESR}_t = \mathrm{ESR} * \frac{N_{\mathrm{s}}}{N_{\mathrm{p}}} \tag{4.17}$$

So

$$\mathrm{d}V = \left(I * \frac{t_{\mathrm{d}}}{C_{\mathrm{t}}}\right) + (I * \mathrm{ESR}_t) \tag{4.18}$$

From Eq. (4.18) t_{d} can be obtained as follows:

$$t_{\mathrm{d}} = C_{\mathrm{t}}\left(\frac{\mathrm{d}V}{I} - \mathrm{ESR}_t\right) \tag{4.19}$$

If the value of t_{d} is lower than a certain value (time required for starting up backup source), then UC cell is replaced by the next size and repeat all previous calculation and so on until the t_{d} is sufficient to start-up backup supply. A computer program is designed for estimating UC sizing with a simple method, Fig. 4.8 displays a flowchart of this program.

4.4.2 Accurate Sizing

UC Values C and ESR depend on voltage and temperature.

1. **Voltage effect**

   The internal structure of the UC is affected by an increased accumulation of charge. As the charge and voltage increase, the effective dielectric constant increases. One possible explanation for this is that the applied voltage extends into the carbon electrode causing a space charge, or an electronic capacitance component to developing [110]. The net result is that UC exhibits a nonlinear

Fig. 4.8 Computer program flowchart for UC simple sizing

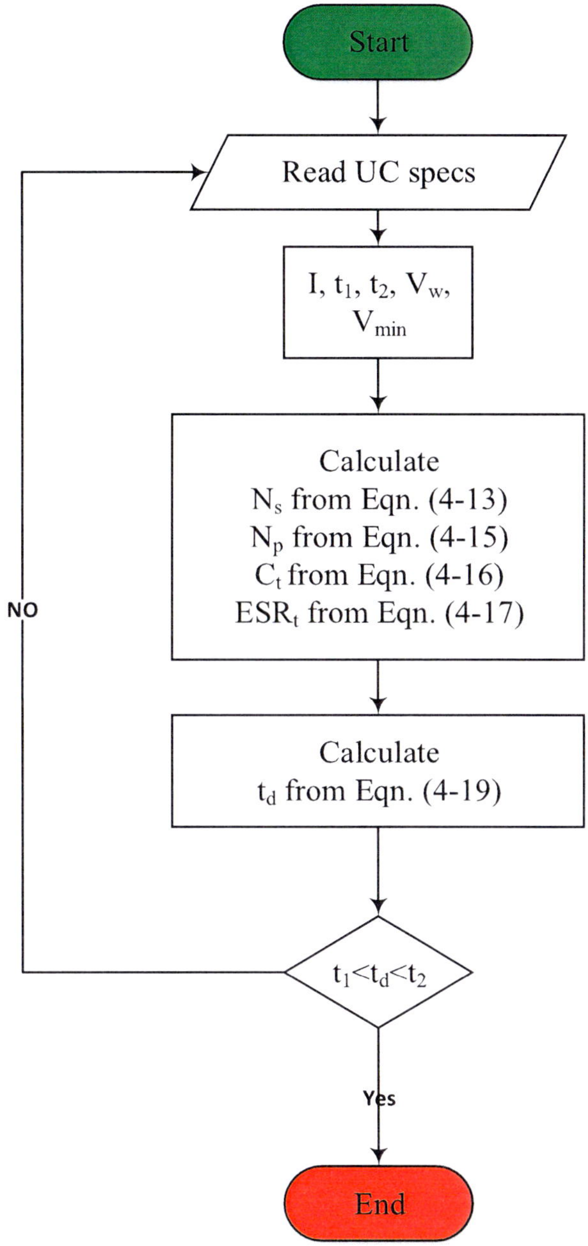

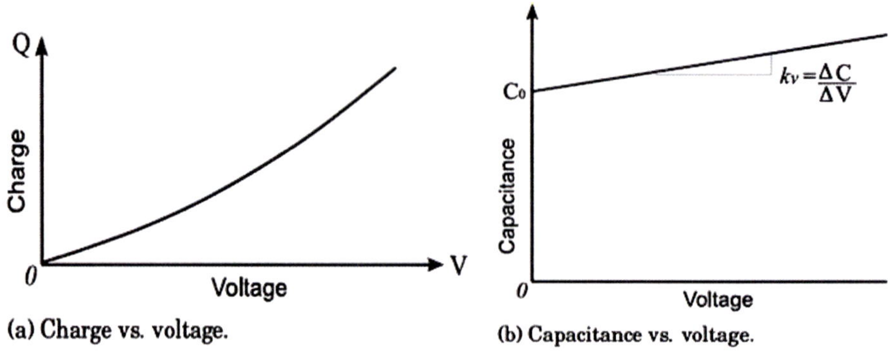

Fig. 4.9 Charging characteristics of UC

charge-voltage characteristic such as that depicted in Fig. 4.9a. Since the capacitance is defined as:

$$C = Q/V \tag{4.20}$$

The capacitance characteristic is also nonlinear. A typical capacitance characteristic of UC is depicted in Fig. 4.9b. At zero voltage, the UC has a base capacitance, C_0, and as the voltage increases, the capacitance increases in an approximately linear fashion.

It should be noted that most manufacturers do not provide sufficient parameters to determine the frequency-dependent effects. Parameters such as the voltage-dependent capacitance and leakage time constant should ideally be determined from experimental results. In many cases, approximations can be used as a starting point. The following estimates were obtained from experimental results presented in the literature [111, 112]: $k_c = 0.1C_{dc}$,

The accurate value of capacitance depends on the value of the voltage between UC terminals so capacitance is [101, 113]:

$$C(v) = C_0 + k_v V \tag{4.21}$$

and

$$C_o = C_{dc} - K_v\, V_{dc} \tag{4.22}$$

where V_{dc}: rated voltage of the capacitor. C_{dc}: capacitance at rated voltage. k_c: voltage-dependent capacitance (F/V).

$$C_{cmin} = C_o + (0.1 * C_o * V_{cmin}) \quad (4.23)$$

$$C_{max} = C_{dc} \quad (4.24)$$

The accurate capacitance value is decreased by 13.73%, and this matches with the specs that states the maximum decrease is limited by 20% of the nominal value [101]. The average capacitance C_{av} can be calculated as follows:

$$C_{av} = (C_{cmin} + C_{max})/2 \quad (4.25)$$

The voltage drops to V_{min} so a number of series UC cells should be:

$$N_s = V_{min}/V_{cmin} \quad (4.26)$$

Temperature Effect

The temperature effect is discussed later. Figure 4.10 and Table 4.2 show the measured capacitance and series resistance for different temperatures for K2 series Maxwell technology [114].

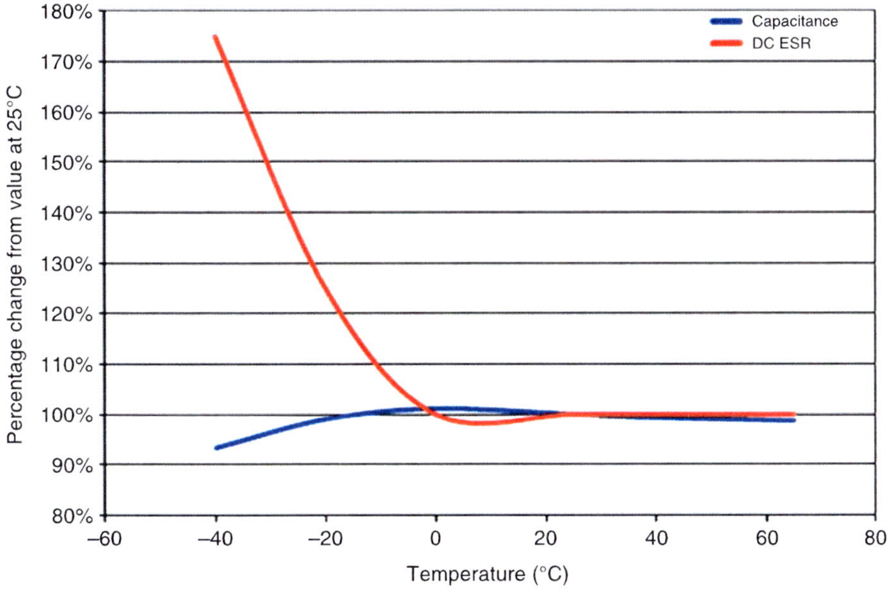

Fig. 4.10 Relationship between operating temperature and UC performance for (K2 series Maxwell technology) UC type

Table 4.2 Experiments results on K2 series Maxwell technology

Item	Temp.	Change due to temp.
$C(T)$	From −40 to 70 °C	No change in capacitance
ESR(T)	From 0 to 70 °C	R
ESR(T)	−10	1.1 R
ESR(T)	−20	1.25 R
ESR(T)	−40	1.75 R

Then complete the sizing with using equations from (4.14) to (4.19). A computer program is designed for estimating UC accurate sizing, Fig. 4.11 displays a flow-chart of this program.

Percentage error between simple sizing method and accurate sizing method can be calculated as:

$$\%\text{age error} = \frac{\Delta t_{\text{d.accur}} - \Delta t_{\text{d.sim}}}{\Delta t_{\text{d.accur}}} \times 100\% \tag{4.27}$$

4.4.3 UC Sizing Using Artificial Neural Networks

Artificial neural networks are used to determine the suitable type of UC. The inputs for ANN are temperature, load current, and duration time. The output is the suitable of UC for inputs.

4.5 Case Study and Applications

4.5.1 Results of Simple Sizing

Ultracapacitors with specs in a Table 4.3 [115] will be used in sizing standby storage system to feed load in time between the stop of main supply until secondary supply comes to in service. In simple sizing, the nominal values of UC are used. The main function of UCs are feeding load after the main supply falls down and start-up standby source.

Almost telecommunication devices and stations have rated voltage 48 V DC with minimum voltage 40 V DC and maximum voltage change from 56 V DC to 72 V DC. In this search $V_{\text{min}} = 40$ V and $V_{\text{max}} = 60$ V, $t_{\text{d}} = 20$ s and current loads 20, 40, 60, 80, 100 A and each value of this current depends on kind of telecom stations and their devices. Table 4.4 and Fig. 4.12 display the results of UC simple size at different loads.

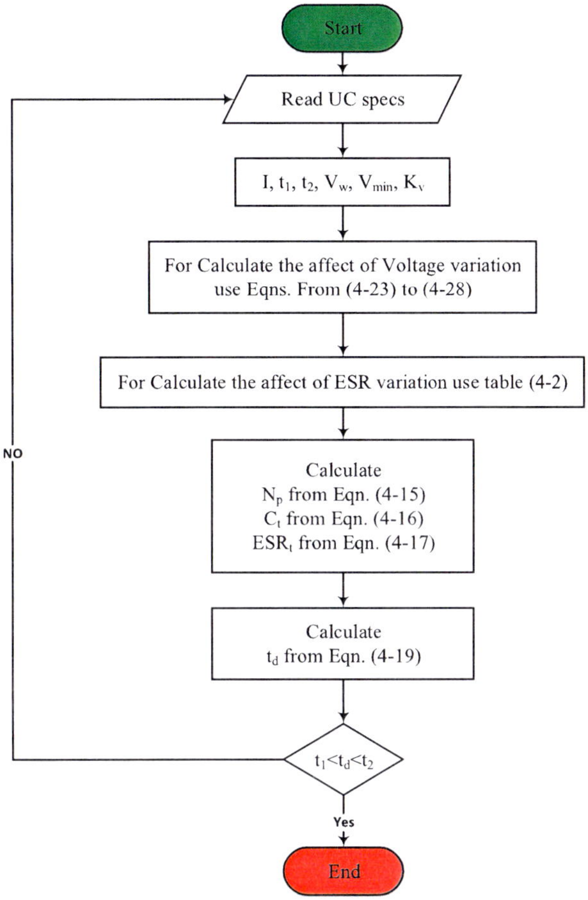

Fig. 4.11 Computer program flowchart for UC accurate sizing

4.5.2 Results of Accurate Sizing

Different values of capacitance can be gotten from Eqs. (4.21) to (4.24) and using K2 series UC in a Table 4.3. Table 4.5 displays the values of capacitance that represented in: rate (C_{rate}), initial (C_{ini}), minimum (C_{min}), and average (C_{av}) values of capacitance.

Due to voltage change and temperature change, the discharge times are influenced as in tables from 4.6, 4.7, 4.8 and 4.9 for different loads and K2 series UC, also Fig. 4.13 displays time durations at different load currents at (0–70° C).

Table 4.3 Characteristics of the K2 series ultracapacitor

UC device name	BCAP0650	BCAP1200	BCAP1500	BCAP2000	BCAP3000
Crated (F)	650	1200	1500	2000	3000
Cmin (F)	650	1200	1500	2000	3000
Max. ESR_{DC} (mΩ)	0.8	0.58	0.47	0.35	0.29
V_{rate} (V)	2.7	2.7	2.7	2.7	2.7
Absolut max. voltage (V)	2.85	2.85	2.85	2.85	2.85
Max. continuous current at 15 °C (A)	54	70	84	110	130
Max. continuous Current at 40 °C (A)	88	110	140	170	210
Cap. change (%)	+20	+20	+20	+20	+20
ESR change (%)	+100	+100	+100	+100	+100

Table 4.4 Discharge time duration for different UC for simple sizing UC

	BCAP650	BCAP1200	BCAP1500	BCAP2000	BCAP3000
I (A)	t_d (s)	t_d (s)	t_d (s)	t_d (s)	t_d (s)
20	28.14	52.21	65.43	87.382	131.403
40	13.81	25.76	32.36	43.391	65.2664
60	9.033	16.94	21.34	28.694	43.221
80	6.645	12.53	15.83	21.346	32.198
100	5.212	9.89	12.52	16.94	25.585

4.5.3 *Comparative Study Between Simple Sizing and Accurate Sizing of UC*

Durations of discharge in simple sizing are greater than durations of discharge in the accurate sizing. Δt represents the difference in discharge durations between simple sizing and accurate sizing for the same load. Δt is represented as the error in simple sizing. Δt is high value at small load current and low value at large load current. At low capacitance of UC, Δt is small. Table 4.10 and Fig. 4.14 display time difference between simple sizing and accurate sizing of UC.

Error percentages in simple sizing method related to accurate sizing method are estimated from Eq. (4.27) and they displayed in a Table 4.11 and Fig. 4.14. From Table 4.11 and Fig. 4.15 the range of percentage error is between 7.28 and 7.54%.

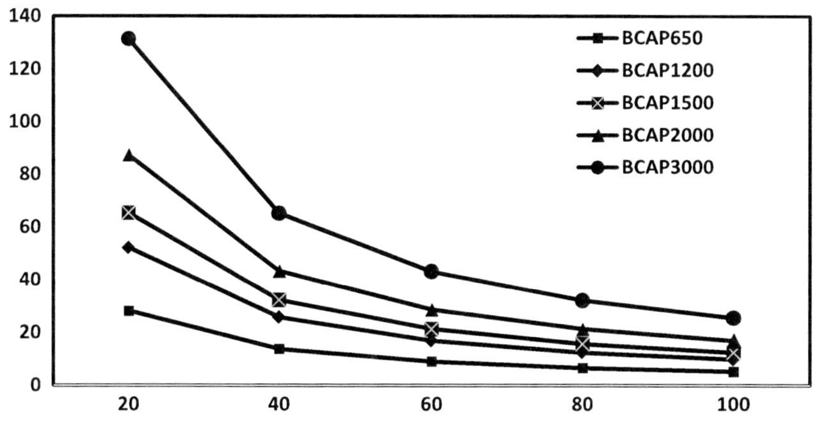

Fig. 4.12 Duration of discharge at different load currents with simple UC sizing

Table 4.5 Capacitance variation due to voltage for K2 series UC

UC	BCAP0650	BCAP1200	BCAP1500	BCAP2000	BCAP3000
C_{rate}	650 F	1200 F	1500 F	2000 F	3000 F
C_{ini}	650 F	1200 F	1500 F	2000 F	3000 F
C_{min}	561 F	1035 F	1294 F	1725	2588 F
C_{av}	605 F	1118 F	1397 F	1863 F	2794 F

Table 4.6 Duration of discharge at 0 °C

	BCAP650	BCAP1200	BCAP1500	BCAP2000	BCAP3000
I (A)	t_d (s)	t_d (s)	t_d (s)	t_d (s)	t_d (s)
20	26.19	48.65	60.94	81.49	122.38
40	12.85	24.00	30.14	40.42	60.78
60	8.4	15.78	19.88	26.73	40.25
80	6.18	11.68	14.74	19.88	29.99
100	4.85	9.21	11.66	15.78	23.83

Table 4.7 Duration of discharge at −10 °C

	BCAP650	BCAP1200	BCAP1500	BCAP2000	BCAP3000
I (A)	t_d (s)	t_d (s)	t_d (s)	t_d (s)	t_d (s)
20	26.14	52.14	65.36	87.41	131.32
40	12.81	25.69	32.29	43.32	65.18
60	8.36	16.87	21.27	28.62	43.13
80	6.14	12.46	15.76	21.28	32.11
100	4.8	9.82	12.45	16.87	25.5

Table 4.8 Duration of discharge at −20 °C

	BCAP650	BCAP1200	BCAP1500	BCAP2000	BCAP3000
I (A)	t_d (s)	t_d (s)	t_d (s)	t_d (s)	t_d (s)
20	26.1	52.04	65.26	87.31	131.19
40	12.7	35.58	32.12	43.22	65
60	8.3	16.77	21.16	28.52	43
80	6.1	12.36	15.65	21.17	31.99
100	4.7	9.71	12.34	16.76	25.37

Table 4.9 Duration of discharge at −40 °C

	BCAP650	BCAP1200	BCAP1500	BCAP2000	BCAP3000
I (A)	t_d (s)	t_d (s)	t_d (s)	t_d (s)	t_d (s)
20	25.83	51.69	64.9	86.96	130.75
40	12.49	25.24	31.83	42.87	64.61
60	8	16.42	20.81	28.17	42.57
80	5.8	12	15.3	20.82	31.54
100	4.5	9.36	11.99	16.41	24.93

Table 4.10 Time difference between simple sizing and accurate sizing of UC

	BCAP650	BCAP1200	BCAP1500	BCAP2000	BCAP3000
I (A)	Δt	Δt	Δt	Δt	Δt
20	1.95	3.56	4.49	5.892	9.023
40	0.96	1.76	2.22	2.971	4.4864
60	0.633	1.16	1.46	1.964	2.971
80	0.465	0.85	1.09	1.466	2.208
100	0.362	0.68	0.86	1.16	1.755

4.5.4 ANN Results for Detecting UC Suitable Type

ANN architecture (3 + 10 + 1) in Fig. 4.16 is used to determine the most suitable type of UC for certain load conditions the classified items of UC types are shown in Table 4.12. Figure 4.17 displays neural network training tool. Also Fig. 4.18 shows validation performance. The training test result are shown in Fig. 4.19. The regression of ANN is shown in Fig. 4.20. The classified items of UC types are shown in Tables 4.12, 4.13 and 4.14.

Figure 4.21 displays the outputs from the analytical method and ANN outputs. It can be noted that there are very small errors between outputs from analytical method and ANN outputs.

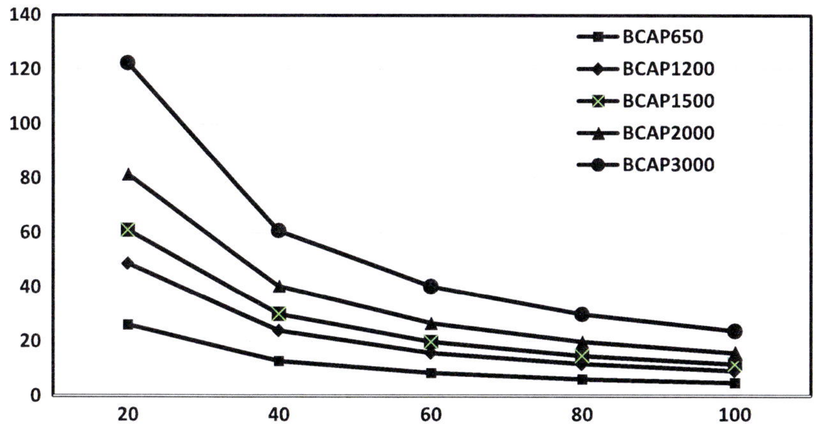

Fig. 4.13 Duration of discharge at different load currents at (0–70°) *C* for K2 series UC

Table 4.11 The percentage error between accurate sizing and simple sizing

I (A)	BCAP650	BCAP1200	BCAP1500	BCAP2000	BCAP3000
20	7.4455899	7.31757451	7.36790286	7.230335	7.37293675
40	7.4708171	7.33333333	7.36562707	7.3503216	7.38137545
60	7.5357143	7.35107731	7.34406439	7.3475496	7.38136646
80	7.5242718	7.27739726	7.39484396	7.3742455	7.36245415
100	7.4639175	7.38327904	7.37564322	7.3510773	7.36466639

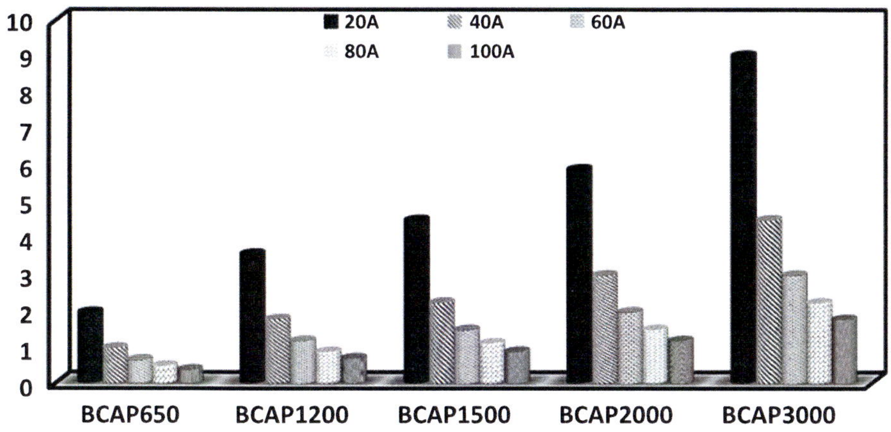

Fig. 4.14 Discharge duration difference between simple sizing and accurate sizing of UC

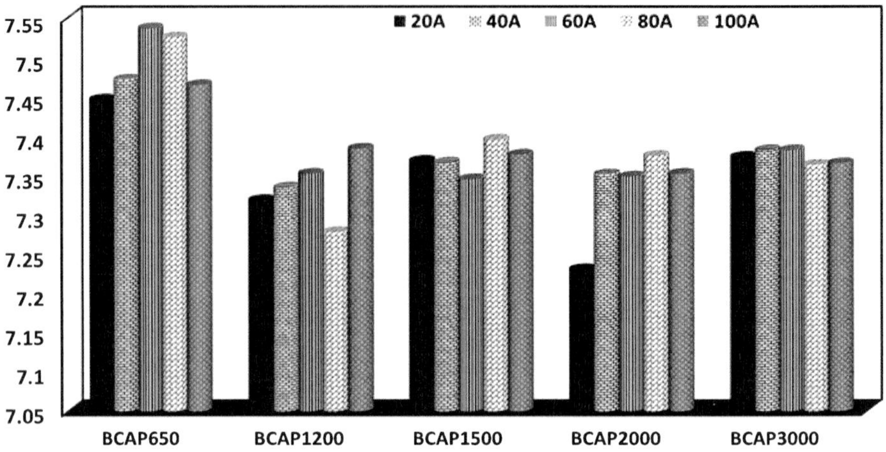

Fig. 4.15 Percentage error between simple sizing and accurate sizing of UC

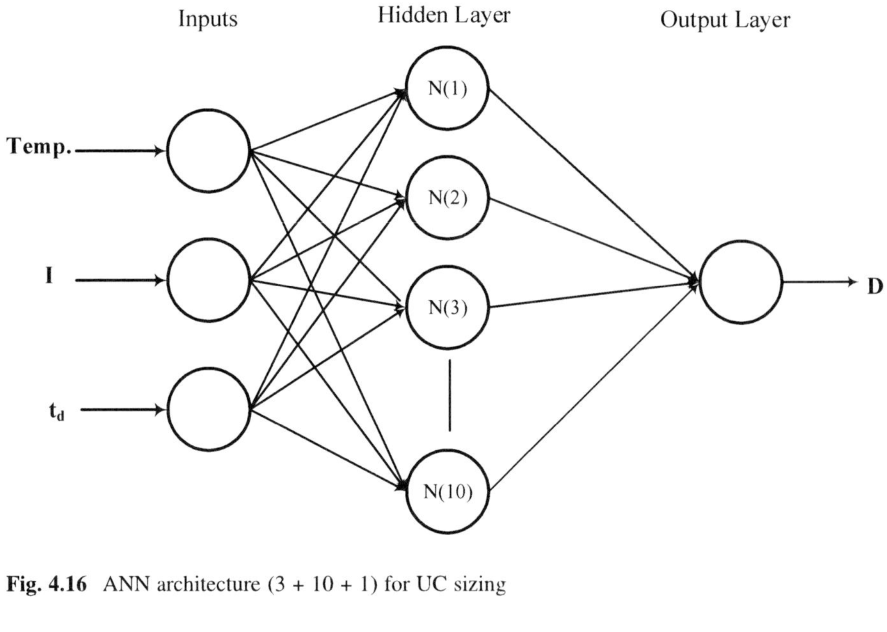

Fig. 4.16 ANN architecture (3 + 10 + 1) for UC sizing

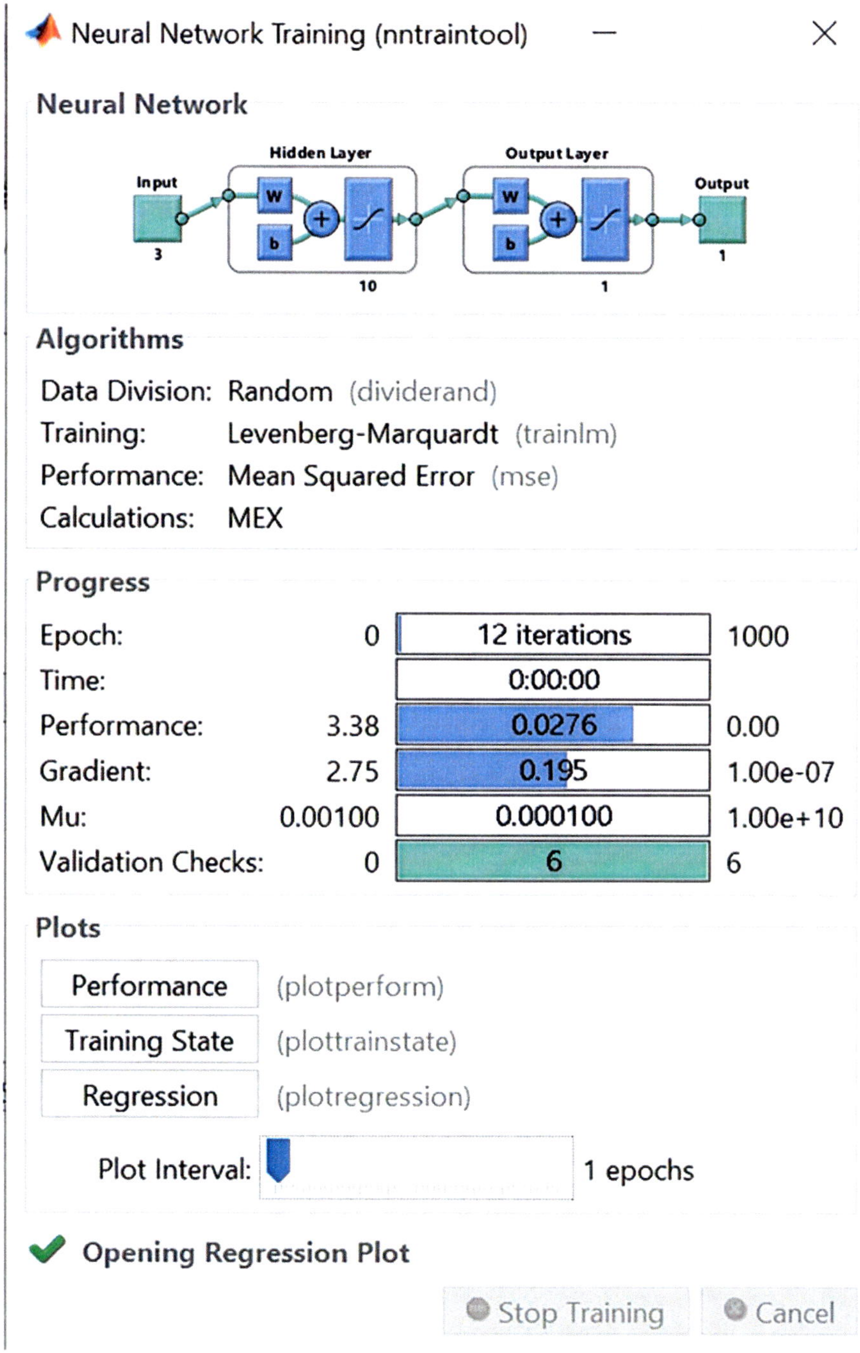

Fig. 4.17 ANN data training

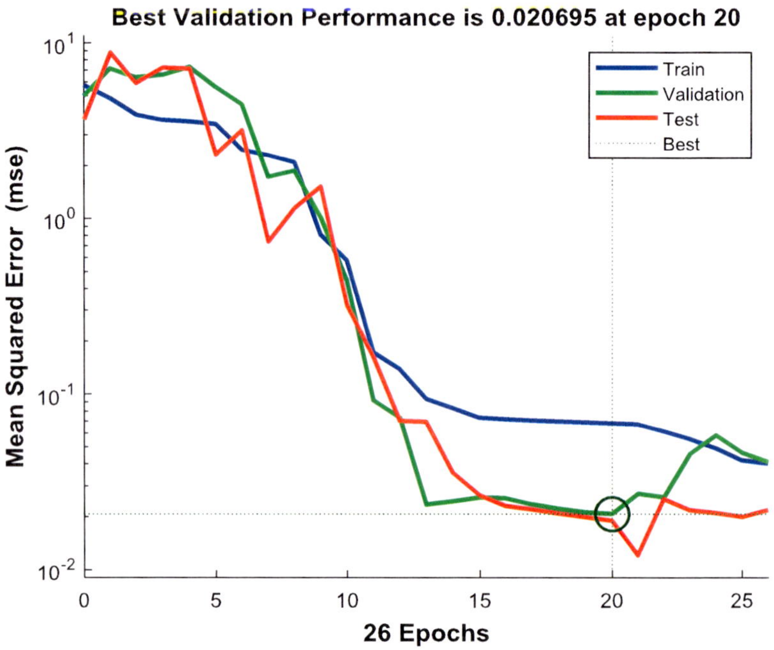

Fig. 4.18 Validation performance for ANN (3 + 10 + 1)

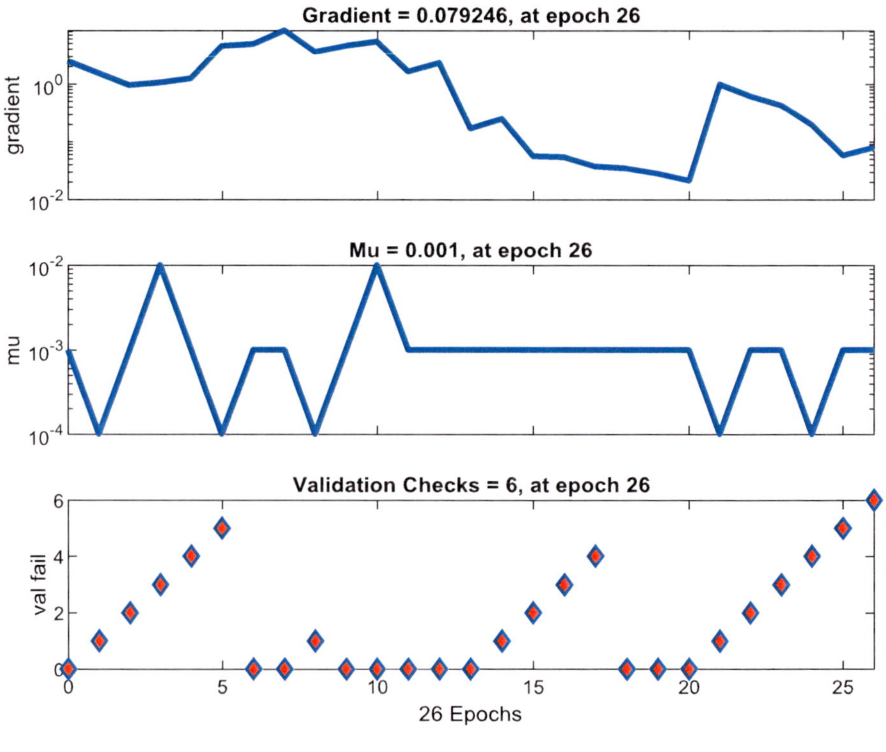

Fig. 4.19 The training test results of ANN (3 + 10 + 1) for UC sizing using ANN

Table 4.12 Classified item of UC types

UC type	Classified item
BCAP650	1
BCAP1200	2
BCAP1500	3
BCAP2000	4
BCAP3000	5

Table 4.13 The weights and bases of the input layer (3 + 10 + 1) ANN for UC sizing

Weights			Base
0.064392	0.25636	1.9535	1.8223
−3.5512	−1.5255	−3.3635	−0.78721
−1.2382	−0.71864	−3.2639	3.8074
1.6101	−0.18779	−1.367	−1.9769
0.17993	3.8407	−0.78785	1.8346
−2.2784	−0.94436	2.5068	2.0845
−0.87805	0.39509	1.991	5.5241
3.0514	1.1732	3.5888	1.3581
−0.53512	0.79774	2.7433	−3.5316
1.7635	0.12289	−2.0214	3.3717

Table 4.14 The weights and bases of the output layer (3 + 10 + 1) ANN for UC sizing

Weights output									
4.603	0.16093	−5.682	−0.09225	1.394	0.25746	6.337	0.28865	2.5951	−0.5644
Bases output									
−0.48763									

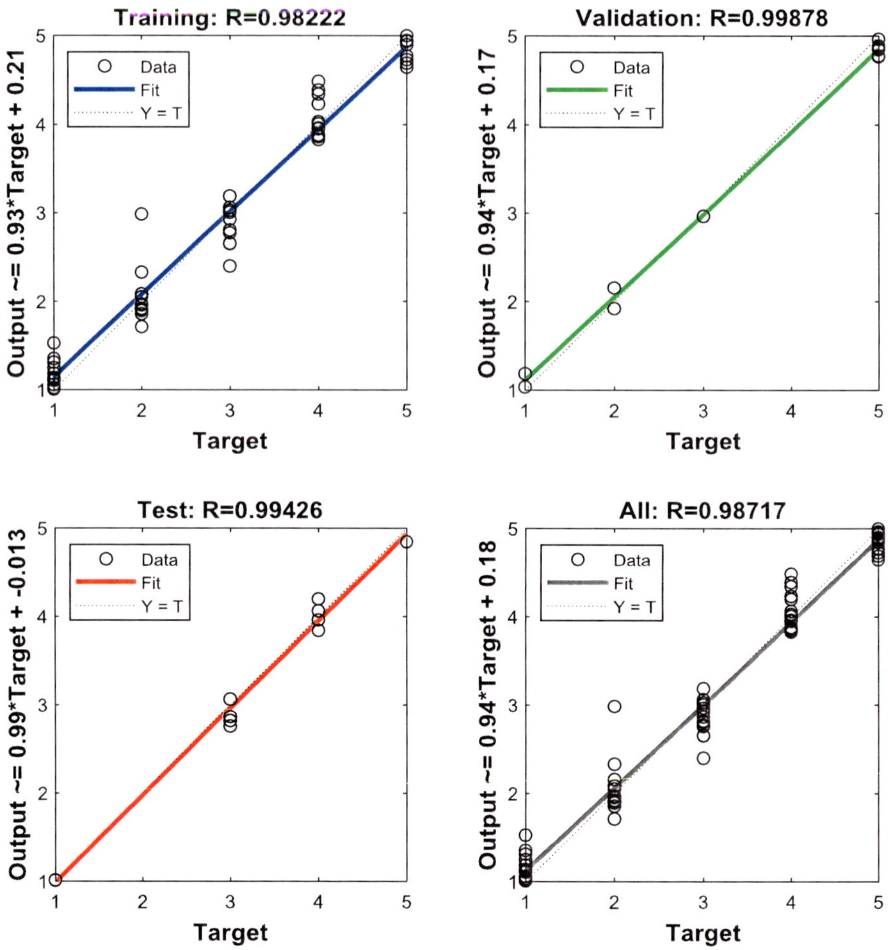

Fig. 4.20 The regression of ANN (3 + 10 + 1) for UC sizing

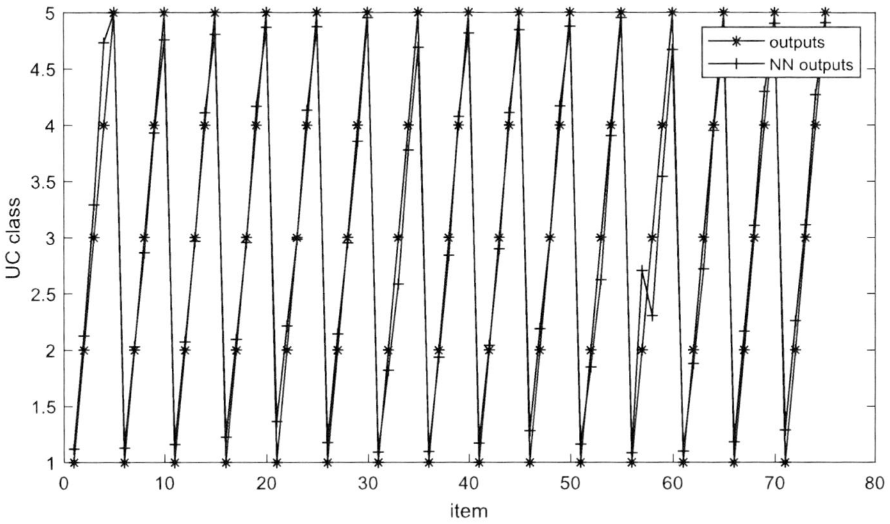

Fig. 4.21 Comparison between results output and NN outputs

4.6 Summary

This chapter exposes ultracapacitors composition, theory, and characteristics. Also, advantages, the equivalent circuits, temperature effect, and voltage variation effect of ultracapacitors are displayed. This chapter also addressed the methods of ultracapacitors sizing. ANN technique is used to determine the optimum UC type for a certain load. Application and results of applied sizing methods and ANN techniques on different loads are displayed.

Chapter 5
Design and Sizing of Photovoltaic Power Systems

5.1 Introduction

The proposed photovoltaic power system, PVPS, which include a photovoltaic module as the main source of energy and DRFC as backup supply and tool for energy storage, finally, UC is used for supplying loads at sudden loads and during stating the time of FC. Figure 5.1 displays off-grid PVPS and Fig. 5.2 display on-grid PVPS. In previous chapters, DRFC and UC are discussed; in this chapter let's discuss the PV part.

The intensity of solar radiation absorbed at the earth's surface mainly depends upon (a) insolation at the earth's surface and (b) properties of the earth's surface. Various factors affect the output of a complete solar PV system. Different parameters such as cell temperature, energy conversion efficiency, and the maximum power point tracking affect the output of a single solar cell. The factors which mainly affect the output of a solar panel are orientation and tilt angle of the solar panel [116].

In order to ensure acceptable operation at minimum cost, it will be necessary to determine the correct size. It is noting that the design should be done on meteorological data, solar irradiance, and the exact load profile of consumers over long periods [117].

The solar energy reaching a solar PV array on the earth surface consists of the main beam which is direct radiation, the diffused beam which is direct radiation affected by atmospheric absorption, and the ground reflected beam which is as a result of reflection of the direct beam from the earth surface as shown in Fig. 5.3; from this figure it can be noted that not all the radiation that is released by the sun will reach the solar PV array surface placed on the earth. This makes the orientation of the solar PV array very important for energy absorption. The direction of the solar PV array has two major parameters: the slope and the azimuth. The slope is the angle of tilt with reference to the ground horizontal surface and the azimuth is the direction towards which the array surface face.

A. A. Elbaset, S. Ata, *Hybrid Renewable Energy Systems for Remote Telecommunication Stations*, https://doi.org/10.1007/978-3-030-66344-5_5

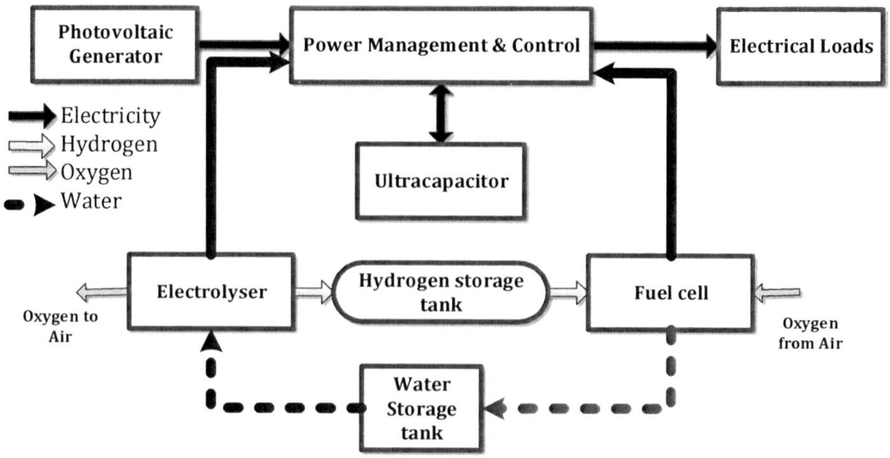

Fig. 5.1 Stand-alone PV/FC/UC power system

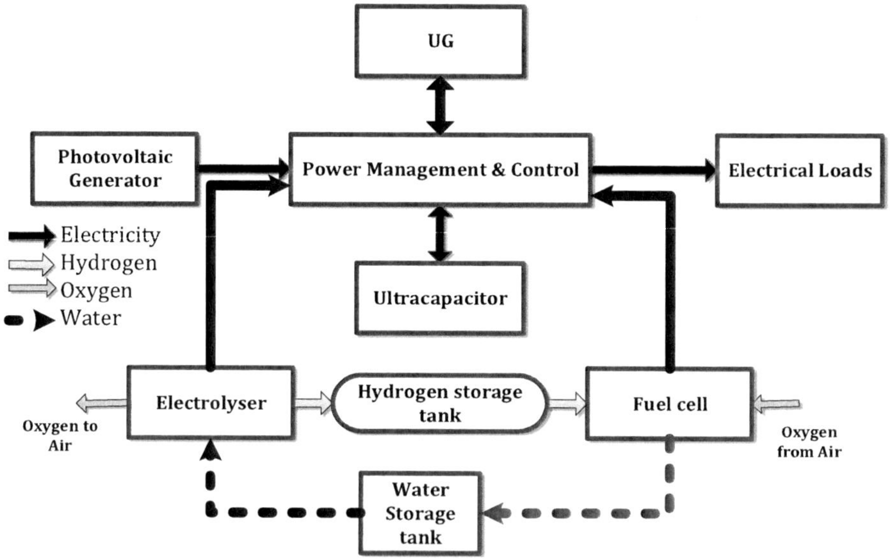

Fig. 5.2 PV/FC/UC power system integrated with UG

When a solar PV array is installed south of the equator, azimuth is due north and when installed north of the equator, azimuth is due south. The azimuth can be due south or due north depending on the latitude of the site or location on the earth's surface. At noon every day, the sun rays are perpendicular to the earth's surface on the equator and give maximum radiation. Any other time of the day, the position of the sun is affected by the latitude, longitude, time of the day, and day of the year. The

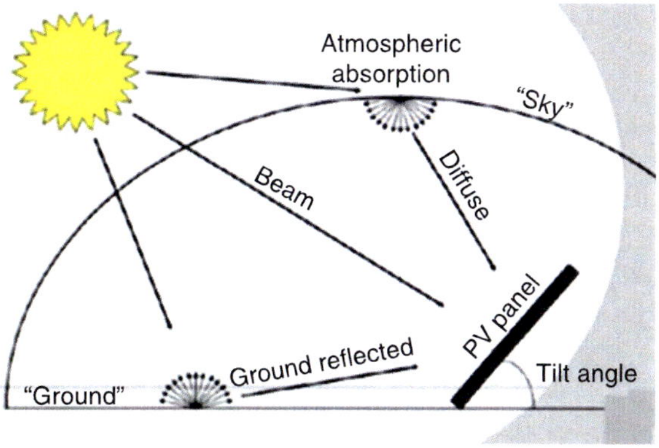

Fig. 5.3 The solar insolation on the PV panel

angle formed between the plane of the equator and a line drawn from the center of the earth is called the solar declination angle denoted by δ.

Global radiation comprises three components:

- Direct solar radiation: The sun radiation received directly from the sun.
- Diffuse radiation scattered by the atmosphere and clouds.
- Reflected radiation from the ground.

The techniques proposed by researchers for calculating optimum tilt angle are usually based on:

(a) Geographic factor $\overline{R}_{\mathrm{b}}$
(b) Clearness index $\overline{K}_{\mathrm{T}}$
(c) Declination angle (δ).

Through experimental analysis, it is proven that the geographic factor method of estimating the optimum tilt angle is the best method for finding the optimum tilt angle of solar panels at a location [118]. This chapter utilizes the geographic factor method for estimating the daily beam radiation on tilted surfaces. The parameters used in determining global solar radiation are as follows:

Declination Angle (δ): The angle between the sun-earth center line and the projection of this line on the equatorial plane. Declinations north of the equator are positive, and those south are negative. The declination angle could be estimated for any given day n using Eq. (5.1).

$$\delta = 23.45 * \sin\ (360 * (284 + n)/365) \tag{5.1}$$

where n is the day number of the year (1–365).

Solar Altitude (α): The angle between a horizontal plane and line joining the centers of the earth and the sun (solar elevation).

Solar Azimuth Angle (Z): The angle between the projection of the straight line joining the centers of the earth and the sun on the horizontal plane and the north-south line at a given location.

Zenith Angle (θ_z): The angle between the sun's rays and the vertical.

Local Latitude (Φ): The angle between a line from the center of the earth to site of interest and the equatorial plane.

Incidence Angle (θ): The angle between the sun's rays and the normal on a surface.

Reflectance (ρ): The ratio of radiation reflected from a surface to that incident on the surface. Reflectivity is the property of reflecting radiation, possessed by all materials to varying extents, called the albedo in atmospheric references. Depending on the type of the surface (ρ) can be determined from [119].

Tilt angle (β): The angle between the horizontal and solar panel.

Hour Angle (ω_s): The angle between the sun projection on the equatorial plane at a given time and the sun projection on the same plane at solar noon.

The earth rotates one revolution per day corresponding to 360° longitude = 24 h. This implies that the day length is 2 × (ω_s/15).

$$\omega_s = \cos^{-1}(-\tan\phi\ \tan\delta) \tag{5.2}$$

where ϕ is the latitude of the location, hence the sunrise and sunset angles for a tilted surface (ω'_s) facing the equator is given by:

$$\omega'_s = \cos^{-1}(-\tan(\phi-\beta)\ \tan\delta) \tag{5.3}$$

Irradiance: The rate at which radiant energy is incident on a surface, per unit area of the surface (W/m^2).

Irradiation: The energy collected per square meter during a specific time interval. If the considered time interval is a day or a year, the terms "daily irradiation" or "annual irradiation" may be used. It has a unit (Wh/m^2).

Beam Irradiation for the Horizontal Surface ($\overline{H}_o$): The monthly mean horizontal daily total extraterrestrial solar radiation energy given by [120]:

$$\overline{H}_o = \frac{24}{\pi} S\left\{1 + 0.033 * \cos\left(\frac{2\pi n}{365}\right)\right\} * (\cos\phi\cos\delta\sin\omega_s + \omega_s\sin\phi\ \sin\delta) \tag{5.4}$$

where n is the day of the year and S is the solar constant equal to 1367 W/m^2.

$$\omega_o = \min\left(\omega_s, \omega'_s\right) \tag{5.5}$$

Clearness Index (K_T): The ratio of the monthly mean horizontal daily radiation to the monthly mean horizontal daily total extraterrestrial solar radiation.

$$\overline{K}_T = \frac{\overline{H}}{\overline{H}_o} \tag{5.6}$$

H is the monthly mean horizontal daily terrestrial solar radiation.

Diffused Irradiation (H_d): The monthly average daily value for the diffused irradiation incident on a horizontal surface.

$$\overline{H}_d = \overline{H} \times \left(1 - 1.13\overline{K}_T\right) \tag{5.7}$$

5.2 Estimation of Total Solar Radiation on an Inclined Surface

Monthly average of total daily radiation on a tilted surface (H_T) can be calculated by calculating the direct beam radiation (H_B), diffuse radiation (H_S), and reflected components (H_R) of the radiation. Thus the total incident radiation on tilted surface is given by the equation:

$$\overline{\mathrm{H}}_{\mathrm{T}} = \overline{H}_{\mathrm{B}} + \overline{H}_{\mathrm{S}} + \overline{H}_{\mathrm{R}} \tag{5.8}$$

5.2.1 *Estimation of Daily Beam Radiation ($\overline{H}_B$)*

The expression for the daily beam radiation received by a tilted solar collector is:

$$\overline{H}_{\mathrm{B}} = \left(\overline{H} - \overline{H}_{\mathrm{d}}\right)\overline{R}_{\mathrm{b}} \tag{5.9}$$

where $\overline{H}$ is monthly average daily global radiation for the horizontal surface.$\overline{H}_d$is the diffuse radiation component for the horizontal surface.$\overline{R}_b$ is the ratio of average daily beam radiation on a tilted surface to that on a horizontal surface [121].

R_b can be estimated by assuming that there is no atmosphere. In the northern hemisphere, the surfaces which are sloped towards the equator, R_b is given as [122]:

$$\overline{R}_{\mathrm{b}} = \frac{\cos\left(\phi - \beta\right)\cos\delta\sin\omega_{\mathrm{o}} + \omega_{\mathrm{o}}\sin\left(\phi - \beta\right)\ \sin\delta}{\cos\phi\cos\delta\sin\omega_s + \omega_s\sin\phi\ \sin\delta} \tag{5.10}$$

where $\omega_o = \min(\omega_s, \omega'_s) = \min(\cos^{-1}(-\tan\phi\tan\delta), \cos^{-1}(-\tan(\phi + \beta)\tan\delta))$.

5.2.2 *Estimation of Ground Reflection Radiation ($\overline{H}_R$)*

The ground reflected radiation can be given after assuming the reflection as isotropic. Hence the daily ground reflected radiation can be expressed as [121, 122]:

$$\overline{H}_R = \overline{H}\rho\frac{(1 - \cos\beta)}{2} \tag{5.11}$$

5.2.3 Estimation of Diffused Sky Radiation ($\overline{H}_S$)

Diffuse irradiance is defined as the solar radiation which is coming from the skydome except from the direct radiation from the sun and the circumpolar region (three degrees of the sun). Sky diffused radiations are difficult to measure because for measuring diffuse sky radiation the pyranometer should be shaded so that it could not receive direct radiation and radiation from the circumpolar region.

Isotropic and anisotropic models are used to determine the ratio of diffuse solar radiation on a tilted surface to that of a horizontal surface. The isotropic models assume that the diffuse sky radiations are uniformly distributed over the skydome. Hence, the diffuse radiation is dependent on the portion of the skydome which can be seen by the collector. The anisotropic models assume that diffuse sky radiations are a collection of anisotropically distributed diffused radiation components in the circumsolar region (sky near the solar disk) and isotropically distributed diffuse component from the rest of the skydome.

The sky-diffuse radiation can be expressed as:

$$\overline{H}_S = \overline{R}_d \times \overline{H}_d \quad (5.12)$$

Here $\overline{R}_d$ is the ratio of the average daily diffuse radiation on a tilted surface to that on a horizontal surface. The diffuse radiation models are as follows [118]:

$$\overline{R}_d = \frac{(3 + \cos(2\beta))}{4} \quad (5.13)$$

After substituting the values of $\overline{H}_S$, $\overline{H}_B$, and $\overline{H}_R$ in Eq. (5.8), the total solar radiation on a tilted surface can be expressed as:

$$\overline{H}_T = (\overline{H} - \overline{H}_d)\overline{R}_d + \frac{\overline{H}\rho(1 - \cos\beta)}{2} + \overline{H}_d\overline{R}_d \quad (5.14)$$

$$\overline{R} = \left(1 - \frac{\overline{H}_d}{\overline{H}}\right)\overline{R}_d + \frac{1 - cos\ \beta}{2} + (1 - 1.13\overline{K}_T)\overline{R}_d \quad (5.15)$$

This equation will be used for determining the optimum tilt angle while substituting the value of $\overline{R}_d$ for the various models described in Eqs. (5.10)–(5.15), respectively.

5.3 Maximum Power Point Tracking (MPPT) in Photovoltaic Power System

There are two main concepts of maximum power. First, at design, special techniques are used in design as follows in the following section. Other in operation, some method is used to have maximum power during the process as will be discussed.

5.3.1 Optimum Solar cells Area (OSCA) Estimation for PVPS

The photovoltaic power system has an enormous capital cost (Capex), so optimization is used for estimating:

1. The optimum values of SCA or a number of solar cell panels used.
2. Capacity rates of power conditioning devices.
3. The amount of stored energy required in the backup energy system.
4. The power rates of fuses, circuit breaker, wires, and so on.

The annual energy balance is used to estimate the OSCA and capacity stored energy in the backup energy system, so the following methodology is used [84, 122, 123]:

1. Hourly PVPS output power, KWh

$$\mathrm{PVO}(t) = \mathrm{SCA} \times \overline{H}_{\mathrm{t}}(t) \times \eta_{\mathrm{cell}} \times \eta_{\mathrm{pc}} \times \eta_{\mathrm{wire}} \times \frac{\mathrm{VF}}{\mathrm{FS}} \tag{5.16}$$

$$\eta_{\mathrm{cell}}(t) = \eta_{\mathrm{cr}} \times [1 - 0.0062 \times (T_{\mathrm{cell}}(t) - T_{\mathrm{cr}})] \tag{5.17}$$

$$T_{\mathrm{cell}}(t) = T_{\mathrm{a}}(t) + \left(\frac{\mathrm{NOCT} - 20}{800}\right) \times H_{\mathrm{T}}(t) \tag{5.18}$$

where SCA is the initial solar cell area, (m^2). $\overline{H}_{\mathrm{t}}$ is the hourly solar radiation on the tilted surface (kW/m^2). $T_{\mathrm{a}}(t)$ is the hourly ambient temperature (°C). $\eta_{\mathrm{cell}}(t)$ is the hourly efficiency of the solar cell. η_{pc} is the efficiency of the power condition. η_{wire} is the efficiency of the wiring and transmission line. VF is the variation factor that influences solar radiation from year to year, (it is about 0.95). FS is a factor of safety including data accuracy and weather conditions and dust (It is about 1.1). $T_{\mathrm{cell}}(t)$ is the hourly cell temperature. NOCT is Nominal Operating Cell Temperature, (°C).

After calculating PVO for total yearly hours (8760 h), make a comparison between hourly PVO and hourly load power required as follows [84, 124]:

$$\sum_{t=1}^{t=8760} (\mathrm{PVO}(t) - P_{\mathrm{Load}}(t)) = \mathrm{d}E \tag{5.19}$$

- If d$E > 0$ then there is surplus energy, so SCA must be decreased by an incremental area and return calculations.
- If d$E < 0$ then there is deficit energy, so SCA must be increased by an incremental area and return calculations.
- If d$E = 0$ then there is no deficit or surplus energy, so SCA is the optimum area (SCA = OSCA).

5.3.2 *Final Solar Cells Area Required*

OSCA was determined without taking into account the effect of the electrical energy storage devices efficiency. If the efficiency of RFC take into account then Final Solar cells area, FSCA, can be calculated as follow:

$$\text{FSCA} = \frac{P_{\text{Tin}_{\text{el}}} \times \text{OSCA}}{P_{\text{Tsur}}} \tag{5.20}$$

where FSCA is final solar cells area, (m^2) and $P_{\text{Tin}_{\text{el}}}$ is total input power to electrolyzer during the year, (kWh). P_{Tsur} : total surplus power with using OSCA, (kWh).

Then the final solar cells area required can be estimated as:

$$\text{FSCA} = \frac{\text{OSCA}}{\eta_{\text{RFC}}} \tag{5.21}$$

5.4 Maximum Power Point Tracking (MPPT)

It is very important with photovoltaic generation to operate the system at high-power efficiency by ensuring that the system is always working at the peak power point regardless of changes in load and weather conditions. The MPPT system is an electronic device inserted between the PV array and the load [125]. This device comprises two essential components, as illustrated in Fig. 5.4.

A DC-DC switching power converter is used along with an MPPT control algorithm to operate the PV system in such a way it can transfer the maximum capable power to the load. As the solar panel outputs power, its maximum generated power changes with the atmospheric conditions (solar radiation and temperature), and the electrical characteristic of the load may also vary. Thus, the PV array internal impedance rarely matches the load impedance.

5.4.1 *MPPT Algorithms Classification*

Several MPPT algorithms have been proposed for PV power systems in recent years, to locate the MPP and increase the system efficiency. The MPPT algorithms may be divided into two types, which are indirect control and direct control or "true tracking techniques." In the indirect control techniques, the MPP is calculated either by measuring the voltage and current of the PV array, the solar insolation, or by the use of mathematical functions obtained from empirical data [126]. Therefore, these

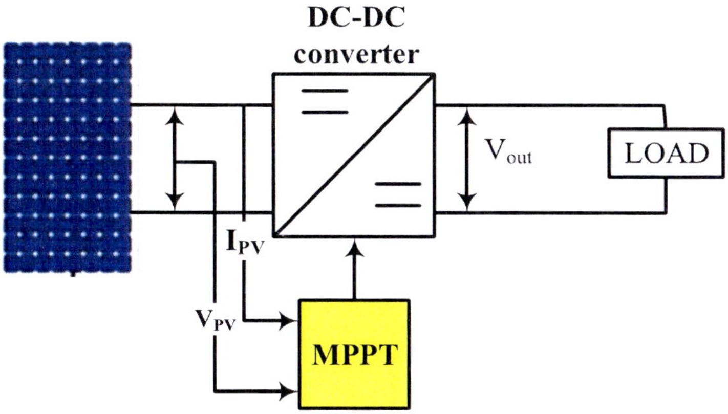

Fig. 5.4 Converter acting as a maximum power point tracker

techniques are incapable of tracking the MPP with varying irradiation and temperature. The indirect control techniques are look-up table technique, constant voltage technique, fractional open-circuit voltage, and fractional short circuit current. However, the direct control techniques can find the optimum operating point even under changing atmospheric conditions, because they do not rely on previous knowledge of or calculated data from the PV array *V-I* characteristics. The actual tracking techniques are the perturbation and observation (P&O) technique, incremental conductance (INC) technique, but artificial intelligence techniques can also be included in this category, such as fuzzy logic control (FLC) techniques and the neural network techniques. In the tracking process, one or two variables are used for calculating the MPP. The fractional open-circuit voltage and fractional short circuit current use only one variable, either the PV array output voltage or current, respectively, though the P&O and INC methods need both variables to determine the MPP.

Neural Networks

NN is called the soft computing method. The input variables can be parameters of the PV array such as V_{oc} and I_{sc}, atmospheric data as irradiation and temperature, or a combination of these. The output is usually the duty cycle for the converter. The performance of the NN depends on the functions used by the hidden layer and how well the neural network has been trained. The links between the nodes are all weighted. The weights are adjusted in the training process. To execute this training process, data of the patterns between inputs and outputs of the NN are recorded over a lengthy period of time, so that the MPP can be tracked accurately.

5.5 On-Grid Photovoltaic Power Systems

The grid-connected PV systems are being developed at a high-speed rate and will soon play a significant role in power electricity generation in some areas. Figure 5.5 displays grid-connected PV systems. Each grid-connected concept consists of a series of paralleled PV panels or strings, and they are configured by a couple of power electronics converters (DC-DC converters and DC-AC inverters) in accordance with the output voltage of the PV panels as well as the power rating.

A significant drawback of the single-stage PV topologies is that the output voltage range of the PV panels/strings is limited, especially in the low power applications (e.g., AC-module inverters), which thus will affect the overall efficiency. The double-stage PV technology can solve this issue since it consists of a DC-DC converter that is responsible for conditioning the voltage of the PV module to a desirable level for the inverter stage. Figures 5.6 and 5.7 show the general block

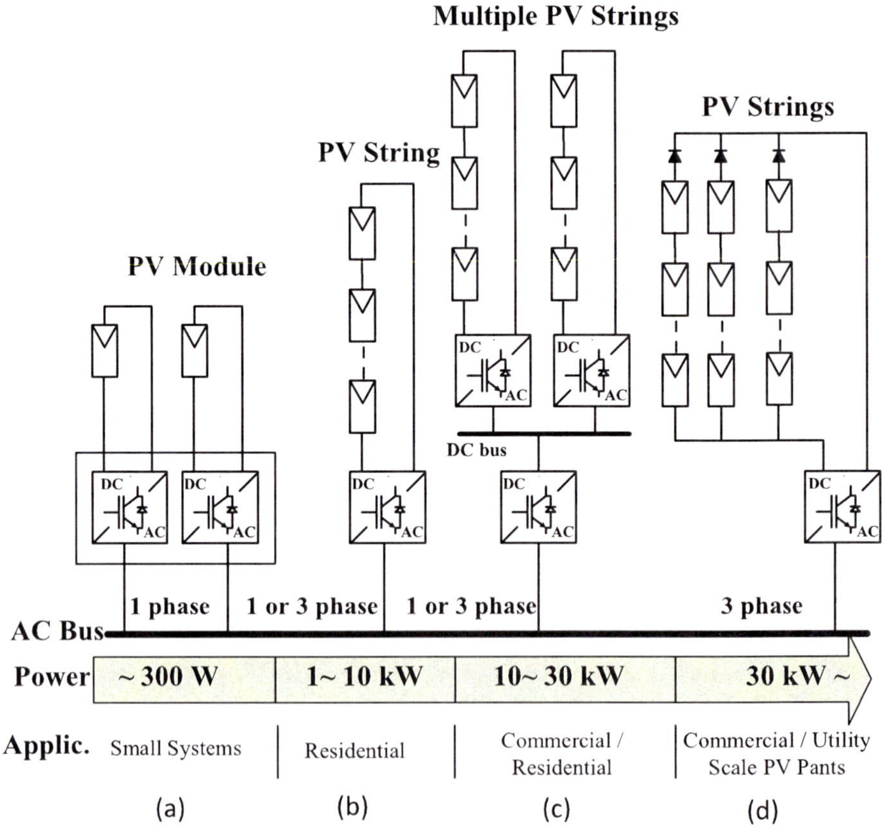

Fig. 5.5 Grid-connected PV system concepts for: (**a**) small systems, (**b**) residential, (**c**) commercial/residential, and (**d**) commercial/utility-scale PV plants

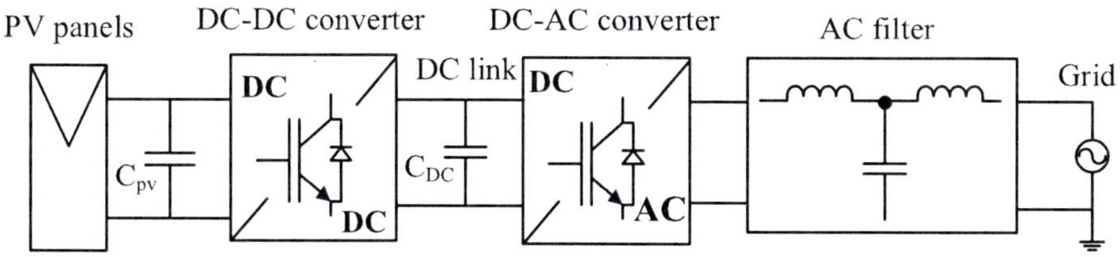

Fig. 5.6 General block diagram of a double-stage single-phase PV topology (with a DC-DC converter)

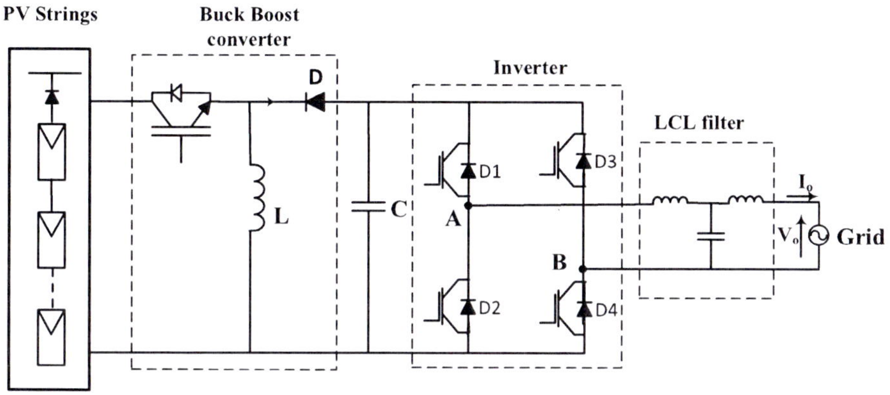

Fig. 5.7 Double-stage single-phase PV topology

diagram of a double-stage single-phase PV topology. The DC-DC converter also performs the MPPT control of the PV panels, and thus, extended operating hours can be achieved in a double-stage PV system. The DC-link capacitor [131] C_{DC} is used for power decoupling, while the PV capacitor C_{pv} is used for filtering.

In general, the DC-DC converter can be included between the PV panels and the DC-AC inverters. Since the power of a single PV module is relatively low and is strongly dependent on the ambient conditions (i.e., solar irradiance and ambient temperature), the trend for AC-module inverters is to integrate either a boost or a buck-boost converter into a full-bridge (FB) or half-bridge (HB) inverter in order to achieve an acceptable DC-link voltage. As it is presented in [130], as shown in Fig. 5.8 a single-stage module-integrated PV converter can operate in a buck, boost, or buck-boost mode with a wide range of PV panel output voltages. An LCL filter is used to achieve a satisfactory THD of the injected current to the grid.

5.5.1 DC-DC Converters

DC-DC converters can be used as switching-mode regulators to convert to dc voltage, usually unregulated, to a regulated dc output voltage. The regulation is

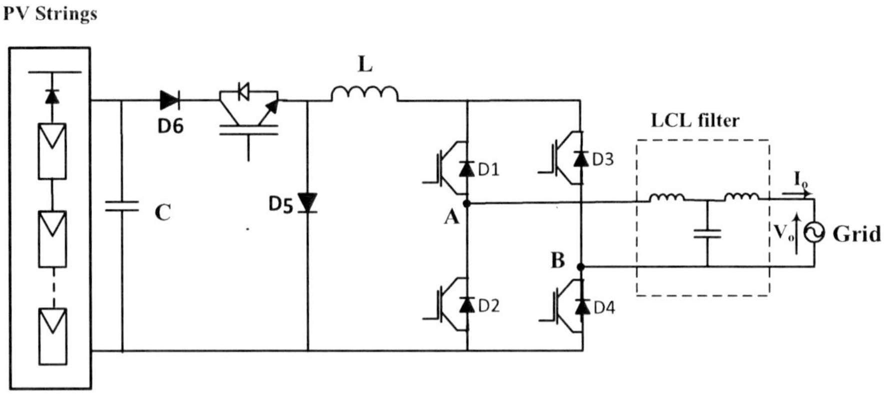

Fig. 5.8 A single-stage grid connected with an LCL filter

normally achieved by PWM at a fixed frequency, and the switching device is normally IGBT or MOSFET [124].

As in AC applications, voltage control devices are needed to step-up and step-down output voltage. The DC-DC converter can be considered as the DC equivalent to an AC transformer with a continuously variable *turns*-ratio. DC-DC converters are very important in all voltage levels applications, like (as), high, or medium, or low power applications. They are much more efficient than the old way of conversion with transformers and designed to deliver an output within any desired range. Moreover, DC-DC converters are extensively used in control of electrical drives; one such application is the regenerative braking of DC drives, where energy is returned into the supply [134].

Functions of DC-DC Converters

The DC-DC converter is designed to function as follows:

1. Convert a DC input voltage into a different value of a DC output voltage.
2. Regulate the DC output voltage depending on the load and line variations.
3. Provide speed control for many applications by controlling the frequency.
4. Provide protection between the input source and the load.
5. Provide protection for the supplying system and the load from electromagnetic interference [135].

There are four types of DC-DC converter [124]:

1. The buck (step-down) converter
2. The boost (step-up) converter
3. The buck-boost (step-up or step-down) converter

The topologies and main ideas of these types can be discussed as follows:

The Buck (Step-Down) Converter

In this type of DC-DC converter, the average output voltage V_o produced is always lower than the DC input voltage. The buck converter is used for voltage step-down in PV applications, such as for charging batteries and in water pumping systems. The basic circuit of the step-down converter is as illustrated below in Fig. 5.9 [136]. The diode (d) is used to enhance the output filtering effect and prevent the switch from absorbing or dissipating the inductive energy because this would lead to overheating the switch. In addition to that, there is an inductor and capacitor at the output of the converter which forms a low-pass filter to attenuate the output voltage fluctuation.

At the time duration when the switch S is on as shown in Fig. 5.10, the diode d becomes reverse biased and the input voltage V_d appears across the inductor leading to a linear increase in the inductor current i_L. Besides, the capacitor C is charged in the same cycle [137].

When the switch S is turned off, as shown in Fig. 5.11, the diode d becomes forward biased, and the voltage across the inductor is reversed $V_L = -V_O$. Therefore, current in the inductor freewheel through the diode starts decreasing linearly. In this cycle the capacitor is charged by the inductor energy. The relation between the input

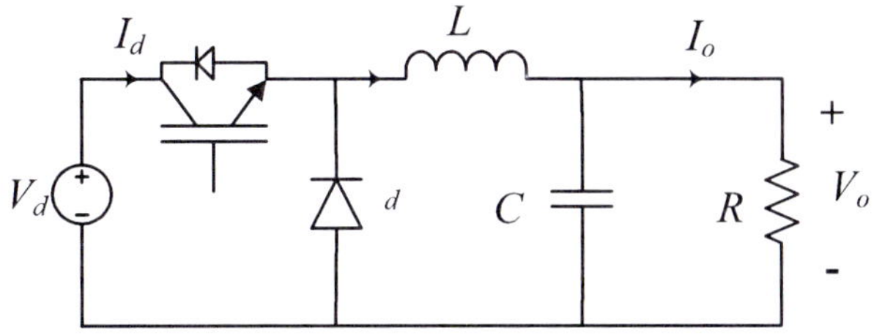

Fig. 5.9 Basic DC-DC Buck converter

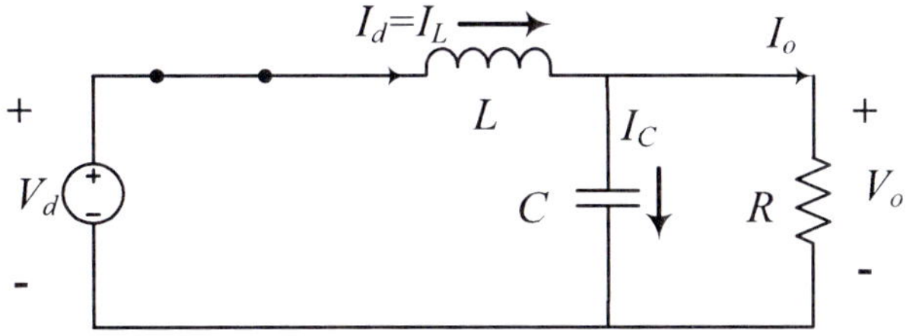

Fig. 5.10 Buck converter switch closed

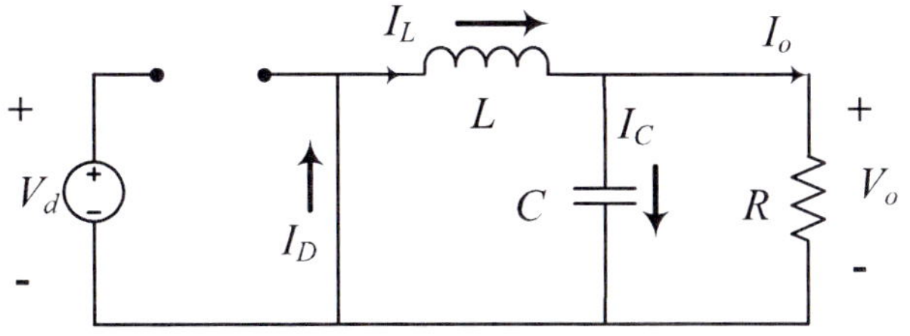

Fig. 5.11 Step-down buck converter switch open

and output voltage, input and output current, and the duty ratio D can be defined by Eqs. (5.22) and (5.23) [138]

$$\frac{V_o}{V_d} = \frac{t_{on}}{T_s} = D \tag{5.22}$$

$$\frac{I_o}{I_d} = \frac{1}{D} \tag{5.23}$$

T_S is the switching period or time period of square pulse that controls the electronics switch S, t_{on} is the on-time of controlling square-wave, I_o is the converter output current, and I_d is the converter input current.

The values of L and C can be estimated as:

$$L = \frac{V_s \times (1 - D)}{2 \times \Delta I_s \times f_s} \tag{5.24}$$

$$C = \frac{\Delta I_o}{8 \times \Delta I_s \times f_s} \tag{5.25}$$

The Boost (Step-Up) Converter

The boost converter can be used in the regenerative braking circuit of DC motors and in regulated DC power supplies. In this type of converter, the output voltage is always higher than the input voltage. Therefore the step-up converter can be applied to MPPT systems where the output voltage needs to be higher than the input voltage, such as in a grid-connected system where the boost converter maintains a high output voltage even if the PV array voltage falls to low values. The circuit topology of the step-up converter is as illustrated in Fig. 5.12 [139, 140]. When the switch S is on, the diode d is reverse biased. Consequently, the current in the inductor L rises linearly due to the input voltage source, and in this case the output stage is isolated and the capacitor C is partially discharged supplying the current load. When the switch is off during the second interval the diode is conducting, and during this time the output stage receives energy from both the inductor and the input source. When

Fig. 5.12 Boost converter

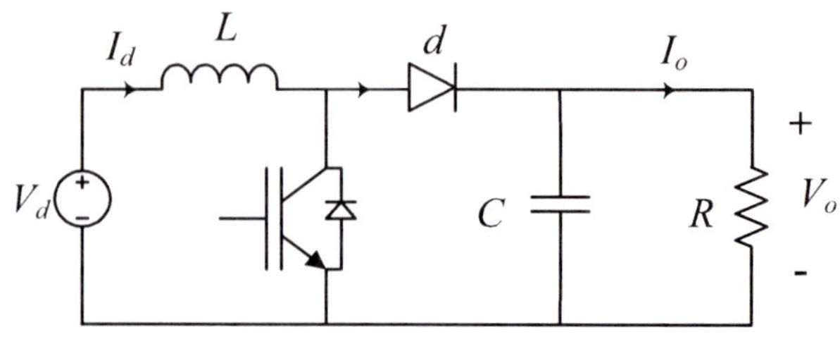

the converter is operating at a steady-state condition, the duty ratio, D, can be expressed as follows [122, 139]:

$$D = 1 - \frac{V_{\mathrm{d}}}{V_{\mathrm{o}}} \tag{5.26}$$

where D denotes the duty ratio, V_{d} and V_{o} denote the input and the output voltages of the converter, respectively. From the above equation, the increase in the duty ratio D will increase the value of the output voltage, V_{o}. Also, the change in the duty ratio results in a change in the input and the output current of the converter. The filter inductor and capacitor to operate the converter in the continuous conduction mode are calculated by the following equations [139]:

$$L = \frac{V_{\mathrm{d}} \times D}{8 \times \Delta I_{\mathrm{L}} \times f_{\mathrm{s}}} \tag{5.27}$$

$$C = \frac{I_{\mathrm{o}} \times D}{\Delta V_{\mathrm{o}} \times f_{\mathrm{s}}} \tag{5.28}$$

Buck-Boost Converter

A buck-boost converter is a cascade connection of two basic converters, the buck converter and boost converter [127]. The buck-boost converter provides an output voltage, which can be higher or lower than the input voltage. In addition, this type of converter produces a negative polarity output with respect to the common terminal of the input voltage. The main application of the buck-boost converter is in regulated DC power supplies. Figure 5.13 shows the circuit topology of the buck-boost converter [140, 141].

In the equivalent circuit when the switch S is turned on during the first time interval t_{on} of the switching period T_{S}, the diode d is reverse based and input provides energy to the inductor causing the inductor current i_{L} to increase. When the switch is off during the second time interval t_{off}, the diode is forward biased, so the energy stored in the inductor is transferred to the load. In this case, the inductor current i_{L} is forced to flow backward through the load and results in a negative polarity in the converter output voltage [140, 142]. During the steady-state

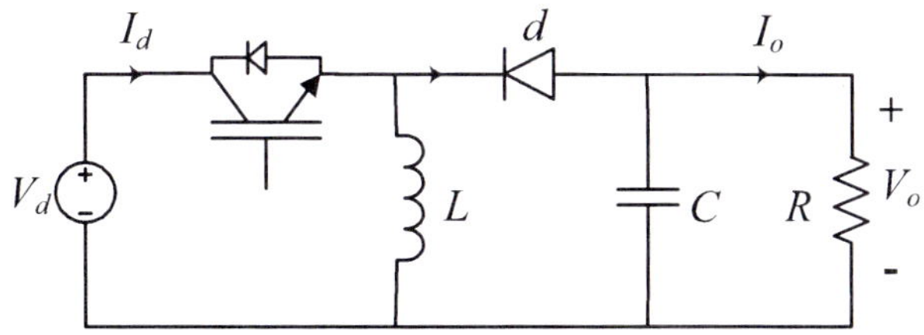

Fig. 5.13 Buck-boost converter

condition, the relationship between the output voltage V_o and the duty ratio of the converter can be defined by the following equation.

$$\frac{V_o}{V_d} = \frac{D}{1 - D} \quad (5.29)$$

From the above equation, the output voltage could be higher or lower than the input voltage according to the value of the duty ratio. For the duty ratio condition $D > 0.5$, the output is higher than the input as in a boost converter. If $D < 0.5$, the output is lower than the input as in a buck converter. So the relationship between D and input/output voltage can be described as follows:

- If $D = 0$, the output is zero.
- If $0 < D < 0.5$, the output voltage is less than the input voltage.
- If $D = 0.5$, the output is equal in magnitude to the input.
- If $0.5 < D < 1$, the output voltage is greater than the input voltage [122].

$$L = \frac{(1 - D)^2 \times R}{2f_s} \quad (5.30)$$

$$C = \frac{D}{R \times f_s \times \left(\Delta V_o / V_o\right)} \quad (5.31)$$

Buck-boost converter is used as a DC-DC converter for all sites to cover the weather conditions all year. The main parameters can be calculated using the following relationships:

Array open-circuit voltage, $V_{oc,a}$ = module open-circuit voltage, V_{oc} * No. of series modules, N_s

$$V_{oc,a} = V_{oc} * N_s \quad (5.32)$$

Array short circuit current, $I_{sc,a}$ = Short circuit current, I_{sc} * No of parallel strings, N_p

$$I_{sc,a} = I_{sc*} N_p \quad (5.33)$$

Array voltage at max. power, $V_{mp,a}$ = module voltage at max. Power, V_{mp}*No. of series modules, N_s

$$V_{mp,a} = V_{mp} * N_s \quad (5.34)$$

Array current at max. power, $I_{mp,a}$ = Short circuit current, I_{mp} * No of Parallel strings, N_p

$$I_{mp,a} = I_{mp*} N_p \quad (5.35)$$

Array max. power transferred, $P_{mp,a}$ = Array voltage at max. power, $V_{mp,a}$ *Array current at max. power, $I_{mp,a}$

$$P_{mp,a} = V_{mp,a} * I_{mp,a} \quad (5.36)$$

Duty cycle, D, can be calculated from Eq. (5.40).
Time period T is:

$$T = 1/f \quad (s) \quad (5.37)$$

where f is the switching frequency of the switches or the MOSFET; it is the frequency of the triangle signal generator/source; f is sufficiently higher than the frequency of the AC components of the circuit voltages and currents [158].

The equivalents resistor, R, can be calculated as follows:

$$R = \frac{V_o^2}{P_o} \quad (5.38)$$

The switching ON and OFF times can be calculated as:

$$t_{on} = D * T \quad (5.39)$$

$$t_{off} = (1 - D)\,T \quad (5.40)$$

Assume the voltage ripple, $(\Delta V_o/V_o)$ is 1%.
Values of L and C can be calculated from Eqs. (5.30) and (5.31).
The relationship between the duty ratio D and output current can be represented as:

$$I_o = \frac{(1 - D)^2 \times V_o}{2 \times f \times L} \quad (5.41)$$

The generated current from PV cell can be calculated as follows [43]:

$$I_{PV} = I_{SC} - I_{rs}\left(e^{\left(q\frac{V_{pv}+IR_S}{aKT}\right)} - 1\right) - \left(\frac{V_{pv} + IR_S}{R_p}\right) \quad (5.42)$$

where

I_{PV}: the photovoltaic current (A)
I_{SC}: the short circuit current (A)
I_{rs}: the reverse saturation current (A)
V_{pv}: the photovoltaic voltage (V)
q: the electron charge (1.6×10^{-19}C)
k: the Boltzmann's constant (1.381×10^{-23} J/K)

T: the junction temperature (K)
a: the diode ideality factor
Rs: the series resistance (Ω)
Rp: the parallel resistance (Ω)

5.5.2 *Inverters*

DC-to-AC converters are known as inverters. The function of the inverter is to change a dc input voltage to symmetric ac output voltage of desired magnitude and frequency [145]. The output voltage could be fixed or variable at a fixed or variable frequency. A variable output voltage can be obtained by varying the input dc voltage and maintaining the gain of inverter constant. The output voltage waveforms of ideal inverters should be sinusoidal [146]. For sinusoidal ac outputs, the magnitude, frequency, and phase should be controllable. Inverters generally use PWM control signals for producing an AC output voltage. According to the type of ac output waveform, these topologies of DC/AC inverters are three typical supply methods [147]:

- Voltage source inverter (VSI), where the independently controlled ac output is a voltage waveform. These structures are the most widely used in small-scale wind power applications.
- Current source inverter (CSI), where the independently controlled ac output is a current waveform. These structures are not widely used in small-scale wind power applications [148].
- Impedance source inverter (z-source inverter or ZSI) is supplied by a voltage source or current source via an x-shaped impedance network formed by two capacitors and two inductors, which is called a z-network. A ZSI has both *buck* and *boost* operation functions. Its output voltage peak value can be higher or lower than the DC-link voltage [149]. Inverters can be broadly classified into two types: single-phase inverters and three-phase inverters [147].

Single-Phase Bridge Inverters

A single-phase bridge voltage source inverter (VSI) is shown in Fig. 5.14. It consists of four switches [145]. When S1 and S2 are turned on, the input voltage V_{d} appears across the load. If S3 and S4 are turned on, the voltage across the load is $-V_{\mathrm{d}}$ [150]:

$$V_{\mathrm{o}} = \sqrt{\left(\frac{2}{T_0}\int_0^{T_0/2} V_{\mathrm{d}}^2 \mathrm{d}t\right)} = V_{\mathrm{d}} \tag{5.43}$$

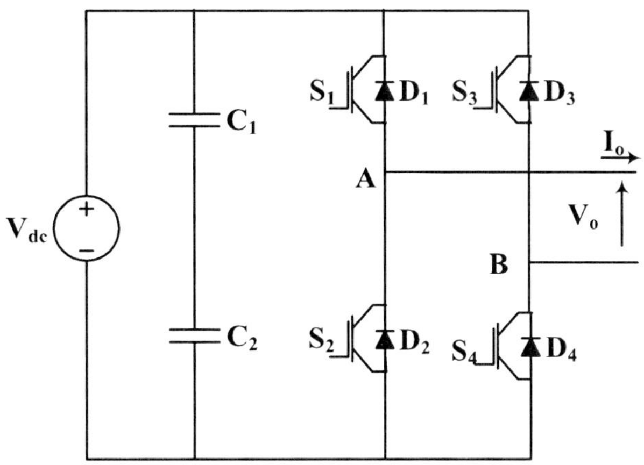

Fig. 5.14 Single-phase full-bridge inverter

The output voltage can be represented in the Fourier series. The RMS value of a fundamental component is:

$$v_o = 0.9\ V_d \quad (5.44)$$

To control the output voltage of inverters, it is often necessary to compensate for the variation of dc input voltage, regulate the output voltage of the inverter, and to adjust the output frequency to the desired value. There are various techniques to vary the inverter gain [145]. The most operational method of controlling the gain and output voltage waveform is sinusoidal pulse-width modulation (SPWM) technique. In the SPWM approach, the width of each pulse is varied in proportion to the amplitude of the sine wave compared at the center of the same pulse [151].

The gating signals are generated by comparing a sinusoidal wave as a reference signal with a triangular carrier wave of frequency, f_c. The frequency of reference signal, f_r determines the output frequency, f_o of the inverter, and the peak amplitude of it A_r specifies the modulation index M. Comparing the bidirectional carrier signal V_{tri} with sinusoidal reference signals $V_{control}$ and $-V_{control}$ results gating signals g_1 and g_4. The output voltage is $V_o = V_d\ (g_1 - g_4)$. However, g_1 and g_4 cannot be released at the same time [151].

Multi-Level Inverters

The voltage source inverters generate output voltage levels either 0 or $\pm V_d$. They are called a two-level inverter. To obtain a quality output voltage or a current waveform with a minimum amount of THD1, they require high-switching frequency and various pulse-width modulation (PWM) techniques. However, switching devices have some limitations in operating at a high frequency such as switching losses and device ratings [152].

Three-Phase Full-Bridge VSI

A three-phase full-bridge VSI is shown in Fig. 5.15. The carrier-based pulse width modulation (PWM) technique is applied in this single-phase full-bridge VSI. Two large capacitors may be used to provide a neutral point *N*, but are not necessary. Six switches S1–S6 in three legs are applied and switched by the PWM signal [153].

5.6 General Control Grid-Connected PV Systems

The control objectives of a single-phase grid-connected PV system can be divided into two major parts: (1) PV-side control with the purpose to maximize the power from PV panels and (2) grid-side control performed on the PV inverters with the purpose of fulfilling the demands to the power grid [155] as shown in Fig. 5.16.

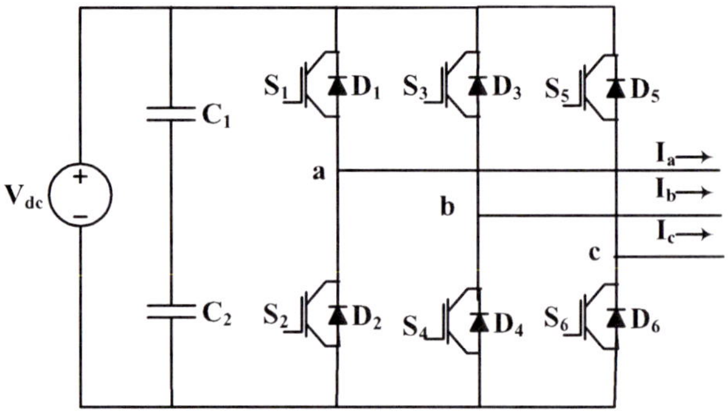

Fig. 5.15 Three-phase full-bridge VSI

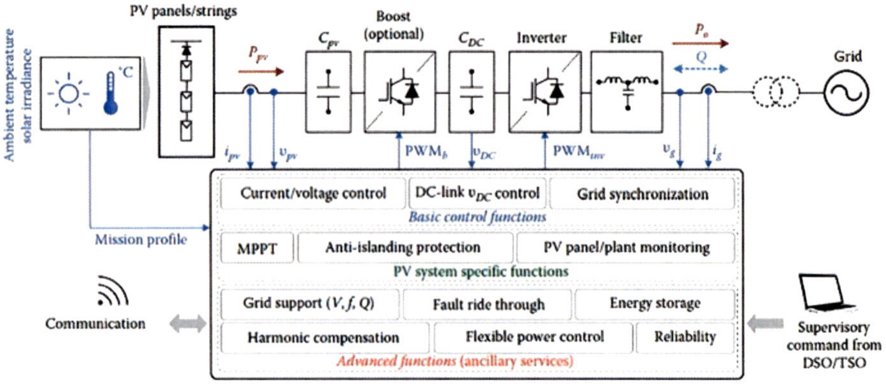

Fig. 5.16 General control blocks (control objectives) of a grid-connected single-phase PV system (*PWM* pulse width modulation, *MPPT* maximum power point tracking) [190]

5.7 Results and Discussions

Communication stations must cover all sites in any country. Any location can have one or more renewable energy sources. In Egypt, almost all sites have solar energy as renewable energy. Other sites have wind energy besides solar energy as sites in the Red Sea coastal and the Mediterranean Sea coastal. So three sites in Egypt are selected to be the application of this study, the first site is Marsa Alam that is located on the southern red sea coast of the eastern desert, Minia is the second site found in Nile Valley, and the third site is Shark El-*Oweinat* found in West desert of Egypt. The latitudes and longitude of the selected location are in Table 5.1.

5.7.1 Estimation of Hourly Radiation on Tilted Surfaces

A proposed computer program has been designed to be dependent upon the methodology in subitem 5.3 to estimate the parameter R and hourly radiation on tilted surfaces. The input data of this program can be summarized as follows:

Radiation and Climate Data

Extraterrestrial radiation and radiation on the horizontal surface are the radiation data required. Also, the average monthly ambient temperature is required. The radiation on the horizontal surface and ambient temperatures have been obtained from [156]. The values of ($\overline{H}_o$) can be estimated from Eq. (5.4). The results of studied cases can be estimated by applying equations from (5.1) to (5.15) and computer program that has flowchart in Fig. 5.17. The following results are gotten, ($\overline{R}$ and $\overline{H}_T$) for best tilt angle and ($\overline{R}_\phi$and $\overline{H}_{T\phi}$) for the three selected sites can be shown in tables from 5.2, 5.3, and 5.4.

5.7.2 Estimation of the Optimum Number of Solar Panels

By applying the technique in Sect. 5.2, and the average hourly loads per month for the selected communication station, Fig. 5.18 shows the required load data and Table 5.5 shows the selected PV panels and the characteristics of them. The main results in this section are optimum solar cells area (OSCA) for each site and each case. Table 5.6 and Fig. 5.19 display OSCA for each location.

Table 5.1 Latitudes and longitudes of the chosen locations

Location	Marsa Alam	Minia	Shark El Oweinat
Latitude	25.0676° N	28.2502° N	22.5857° N
Longitude	34.8790° E	29.9741° E	28.7166° E

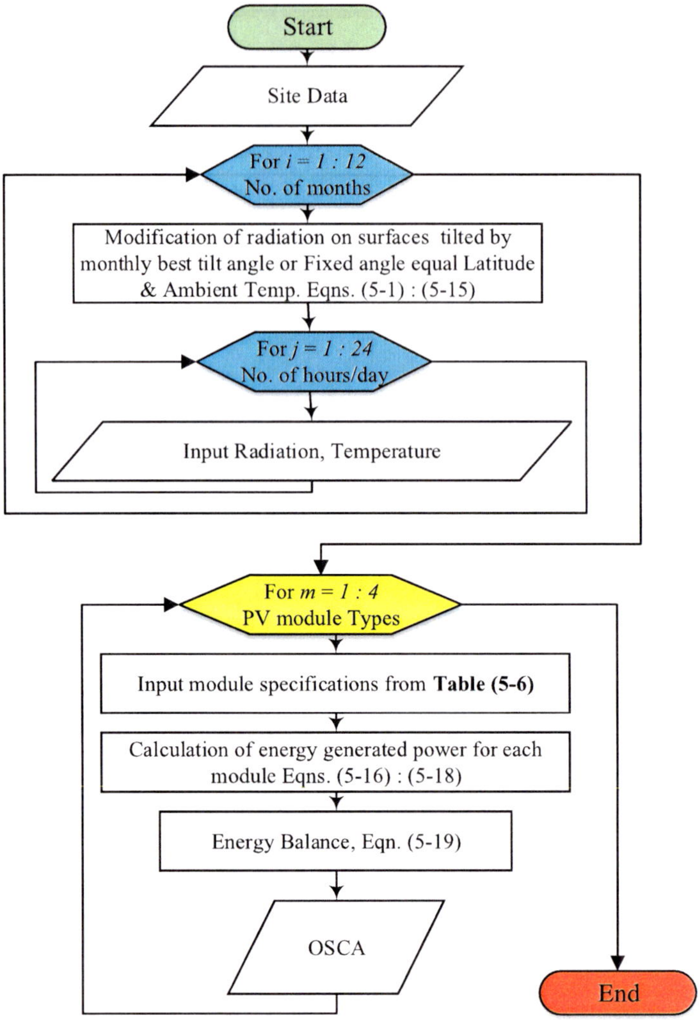

Fig. 5.17 The flowchart of computer program for estimating OSCA

The percentage difference in OSCAs between monthly best tilt angle and latitude angle can be shown in Table 5.7 and Fig. 5.20.

It can be noted from Tables 5.6, 5.7 and Figs. 5.19, 5.20 that:

1. OSCA is inversely proportional to PV module efficiency. (high PV module efficiency means decrease OSCA and vice versa)
2. The monthly best tilt angle gives the lower OSCA from all selected PV modules. Related with latitude angle.

Table 5.2 Design parameters for Marsa Alam site

Date	$\overline{H}$ (KWh/m^2)	$\overline{H}_o$ (KWh/m^2)	$\overline{K}_T$	δ°	Tilt angle (β°)	$\overline{R}$	$\overline{H}_T$	$\overline{R}_\phi$	$\overline{H}_{T\phi}$
17 Jan.	4.1414	6.64	0.623705	−20.9	45.9	1.471	6.09	1.34	5.55
16 Feb.	5.4127	7.8	0.693936	−12.91	37.9	1.294	7.01	1.242	6.72
16 Mar.	6.5059	9.12	0.713366	−2.36	27.4	1.113	7.24	1.111	7.23
15 Apr.	7.1743	10.3	0.696534	9.47	15.5	1.008	7.23	0.986	7.08
15 May	7.3759	10.93	0.674831	18.83	6.2	1	7.38	0.905	6.67
11 Jun.	6.5334	11.12	0.587536	23.099	1.9	1	6.53	0.879	5.74
17 Jul.	7.6776	11.55	0.664727	21.15	3.8	1	7.68	0.886	6.81
16 Aug.	7.2423	10.5	0.689743	13.387	11.6	1	7.24	0.951	6.88
15 Sep.	6.4165	9.51	0.674711	2.1292	22.9	1.058	6.79	1.057	6.78
15 Oct.	5.0007	8.17	0.612081	−9.68	34.7	1.201	6.01	1.177	5.89
14 Nov.	4.7956	6.91	0.694009	−18.97	44	1.456	6.98	1.336	6.41
10 Dec.	4.1698	6.29	0.662925	−23.07	48	1.573	6.56	1.397	5.82

Table 5.3 Design parameters for Shark El-Oweinat site

Date	$\overline{H}$ (KWh/m^2)	$\overline{H}_o$ (KWh/m^2)	$\overline{K}_T$	$\delta°$	Tilt angle ($\beta°$)	$\overline{R}$	$\overline{H}_T$	$\overline{R}_\phi$	$\overline{H}_{T\phi}$
17 Jan.	3.768	7	0.538286	−20.9	43.8	1.367	5.1516	1.261	4.750
16 Feb.	5.661	8.12	0.697167	−12.9	35.81	1.261	7.139	1.21	6.847
16 Mar.	7.256	9.36	0.775214	−2.36	25.26	1.104	8.0137	1.102	7.996
15 Apr.	7.812	10.38	0.752601	9.47	13.43	1.005	7.848	0.983	7.675
15 May	8.002	10.88	0.735478	18.83	4.067	1	8.002	0.903	7.227
11 Jun.	8.638	11.04	0.782428	23.1	−0.199	1	8.638	0.864	7.46
17 Jul.	8.548	10.97	0.779216	21.15	1.75	1	8.548	0.880	7.525
16 Aug.	7.221	10.52	0.686407	13.39	9.513	1	7.221	0.95	6.856
15 Sep.	7.191	9.62	0.747505	2.13	20.771	1.052	7.5679	1.052	7.562
15 Oct.	5.811	8.4	0.691786	−9.7	32.584	1.198	6.9601	1.17	6.796
14 Nov.	4.821	7.24	0.665884	−18.97	41.87	1.392	6.7091	1.281	6.1759
10 Dec.	4.309	6.665	0.646512	−23.07	45.968	1.502	6.4703	1.336	5.755

Table 5.4 Design parameters For Minia Site

Date	$\overline{H}$ (KWh/m^2)	$\overline{H}_o$ (KWh/m^2)	$\overline{K}_T$	$\delta°$	Tilt angle ($\beta°$)	$\overline{R}$	$\overline{H}_T$	$\overline{R}_\phi$	$\overline{H}_{T\phi}$
17 Jan.	2.43	6.16	0.3944	−20.895	48.975	1.3404	3.26	1.2725	3.09
16 Feb.	4.96	7.402	0.6700	−12.91	40.99	1.3399	6.65	1.2879	6.39
16 Mar.	6.71	8.838	0.7592	−2.3597	30.44	1.1519	7.73	1.1493	7.71
15 Apr.	7.72	10.18	0.7583	9.47275	18.607	1.0192	7.87	0.9973	7.7
15 May	7.9	10.96	0.7208	18.8326	9.2474	1	7.9	0.8991	7.1
11 Jun.	8.33	11.22	0.7424	23.0985	4.9815	1	8.33	0.8561	7.13
17 Jul.	7.96	11.06	0.7197	21.15	6.9300	1	7.96	0.8769	6.98
16 Aug.	7.53	10.44	0.7212	13.3867	14.693	1	7.53	0.9538	7.18
15 Sep.	6.77	9.288	0.7288	2.1292	25.951	1.0839	7.34	1.0831	7.33
15 Oct.	5.65	7.804	0.7239	−9.684	37.764	1.2846	7.26	1.2524	7.08
14 Nov.	4.27	6.734	0.6340	−18.97	47.05	1.496	6.39	1.3835	5.91
10 Dec.	3.74	5.808	0.6439	−23.068	51.148	1.6574	6.2	1.4773	5.53

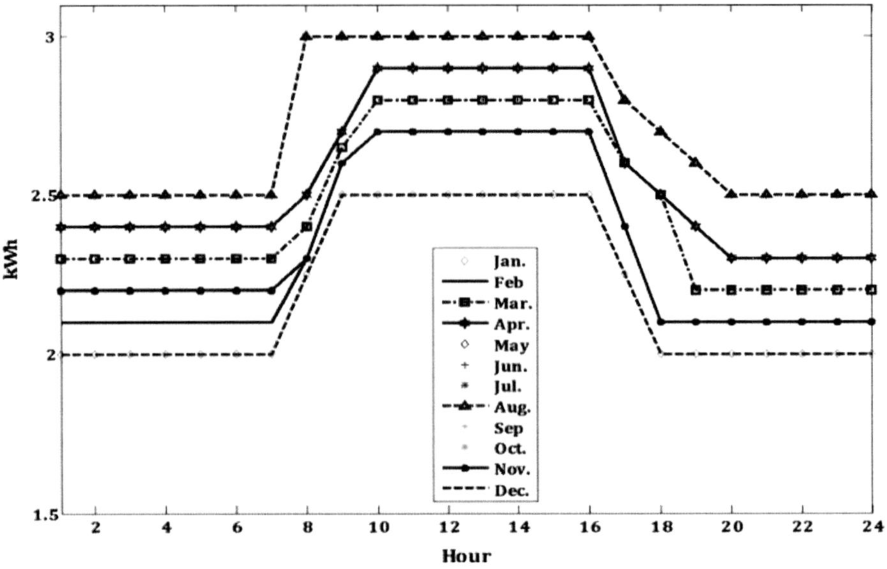

Fig. 5.18 Average hourly electrical loads

Table 5.5 The characteristics of the selected PV modules [157]

PV module				
Item	FVG 185M-MC	YS280-60	RS-M300	SPR-X22-360
Module efficiency (%)	14.6	18.03	18.3	22.2
Maximum power (P_{max}), W	185	280	300	360
Voltage at P_{max} (V_{mp}), V_{dc}	37.5	32.1	32.6	59.1
Current at P_{max} (I_{mp}), A	4.95	8.72	9.19	6.09
Short circuit current (I_{sc}), A	5.4	9.45	9.64	6.48
NOCT, °C	47 ± 2	45 ± 2	44 ± 2	47 ± 2
Maximum system voltage (V_{dc})	700	1000	1000	1000
Module active area, m^2	1.729	1.6335	1.6368	1.629668
Maximum series fuse rating, A	7	15	15	15
Open circuit voltage (V_{oc}), V_{dc}	44.5	39.2	39.9	69.5

From the previous results, SPR-X22-360 solar panel is optimum for all locations, because it gives minimum OSCA required in cases of latitude tilt angle and monthly optimum tilt angle for the selected locations.

Other results from energy balance programs represent: Surplus / Deficit energy and accumulated energy during the year, these results are estimated by using

Table 5.6 Energy balance program results for selected PV modules and monthly best tilt angle and latitude angle

Solar cell type		FVG 185M-MC	YS280-60	RS-M300	SPR-X22-360
Marsa Alam site	OSCA with Best tilt angle, m^2	79.7	61.9	61	50.262
	OSCA with latitude angle, m^2	81.65	65.9	65.1	53.654
Minia site	OSCA with Best tilt angle, m^2	84.3	68.25	67.25	55.4299
	OSCA with latitude angle, m^2	89.42	72.41	71.34	58.805
Shark El-Oweinat Site	OSCA with Best tilt angle, m^2	78.04	63.195	63.1	51.3232
	OSCA with latitude angle, m^2	83.41	67.541	66.545	54.8542

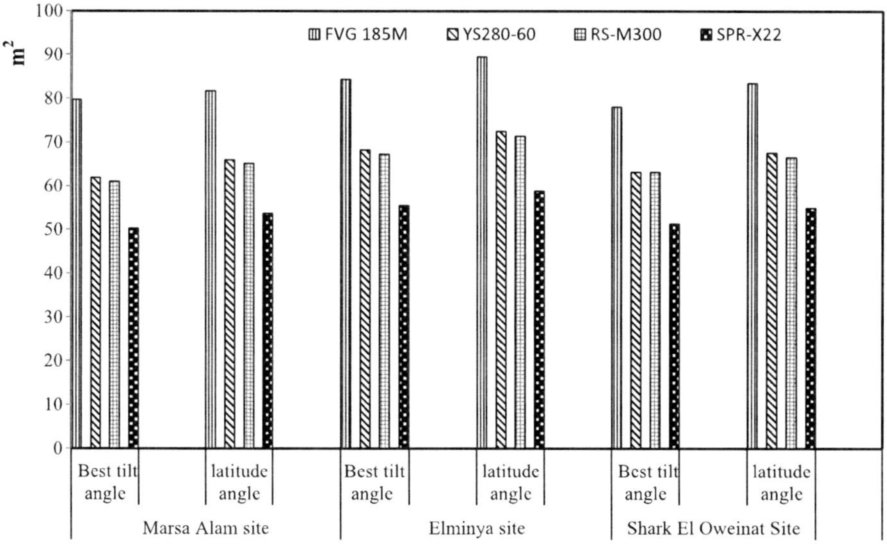

Fig. 5.19 OSCA for selected sites using selected PV modules and monthly best tilt angle and latitude angle

Table 5.7 Percentage difference in OSCAs between the best tilt angle and latitude angle for selected PV modules

PV module	Marsa Alam (%)	Minia site (%)	Shark El Oweinat site (%)
FVG 185M-MC	2.446675	6.0735	6.8811
YS280-60	6.462	6.0952	6.8771
RS-M300	6.7213	6.0818	5.4596
SPR-X22-360	6.7486	6.08895	6.87993

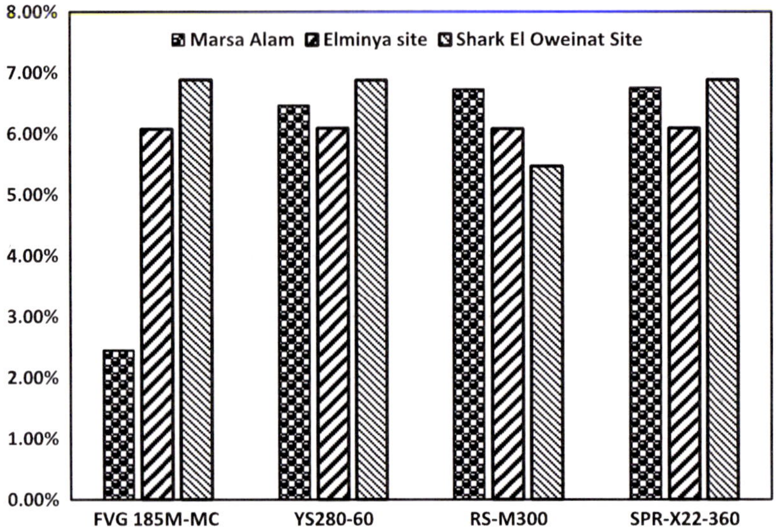

Fig. 5.20 Percentage area differences between the best tilt angle and latitude angle for selected PV modules

SPR-X22-360 photovoltaic module. Tables from 5.8, 5.9, and 5.10 and figures from 5.21, 5.22, and 5.23 display these results.

5.7.3 Rates of Hydrogen Production and Usage in PVPS

Regenerative fuel cells system is used to supply communication station with specific hourly loads in the three selected sites. The hourly loads are discussed in item 5.5, DRFC is a backup power supply the EZ part is HyLYZER^{TM}-2, Table 5.11 displays its technical specs of selected EZ:

The other part of the DRFC is H-3000 (FCS-B3000) that represents the fuel cell part. Its specs are shown in Table 5.12.

A computer program that has a flowchart in Fig. 5.24 is prepared to determine hydrogen production and hydrogen usage. The results are displayed in tables from 5.13, 5.14, and 5.15 and figures from 5.25, 5.26, 5.27, 5.28, 5.29, and 5.30. The same energy generated at both the monthly best tilt angle and latitude angle, whereas a shortage in radiation incident in the latitude angle is substituted by increasing the solar cells area. So hydrogen production due to surplus energy is about the same in two cases of solar panel orientation. The positive values are surplus, and the negative values are deficit.

Table 5.8 Surplus/deficit energy during the year for Marsa Alam, (kWh)

Month	Jan	Feb	Mar	Apr	May	Jun	Jul	Aug	Sep	Oct	Nov	Dec
d*E*	0.174	248	322	130.75	−189	−213	−162	−155	−156.4	−251.9	350.58	75.1

Table 5.9 Surplus/deficit energy during the year for Shark El Oweinat, kWh

Month	Jan	Feb	Mar	Apr	May	Jun	Jul	Aug	Sep	Oct	Nov	Dec
dE	−222.1	197.1	439.2	172	−115	−47.9	−69.2	−214	−13.4	−72	−27.8	−26.5

Table 5.10 Surplus/deficit energy during the year for Minia Site, kWh

Month	Jan	Feb	Mar	Apr	May	Jun	Jul	Aug	Sep	Oct	Nov	Dec
dE	−635.9	197.9	394.5	287.2	−212.9	−142.3	−158.3	−40	79.1	91	33.1	106.4

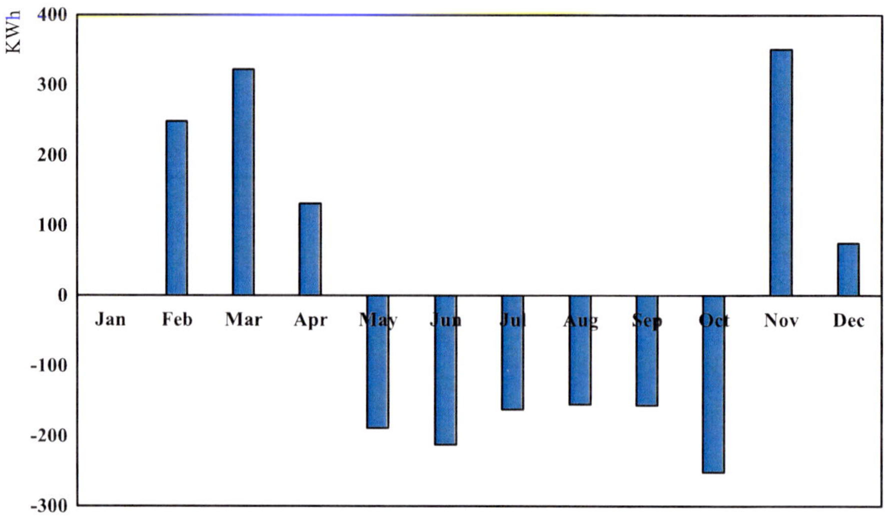

Fig. 5.21 Surplus/deficit energy during the year for Marsa Alam, kWh

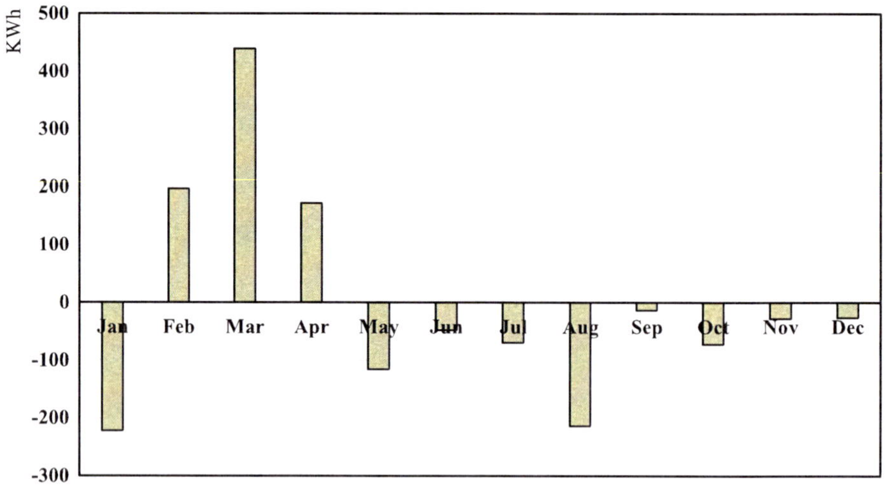

Fig. 5.22 Surplus/deficit energy during the year for Shark El Oweinat, kWh

From Sect. 3.6, volumes of water and hydrogen storage tanks can be estimated. Table 5.16 displays water tank storages at different sites and hydrogen tank storage volumes at different pressure. (1 N m^3 hyd is given from 1-liter water).

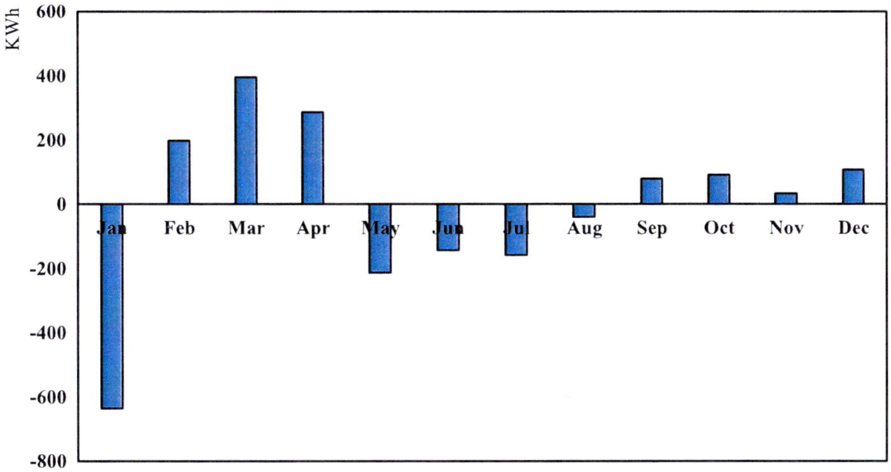

Fig. 5.23 Surplus/deficit energy during the year for Minia, kWh

Table 5.11 Technical specification of HyLYZER™-2 [84]

Model	HyLYZER™-2
Type	PEM electrolyzer
System efficiency	6.7 kWh/Nm^3
Hydrogen output pressure	0–7.9 barg
Hydrogen purity	>99.998%
Power supply	200–260, 1 phase, two-wire + gnd, 50/60 Hz 208/120, 3 phase, four-wire + gnd, 50/60 Hz Direct connection to DC possible upon request

Table 5.12 Technical specification of H-3000 (FCS-B3000) (fuel cell part) [84]

Type	PEM
Number of cells	72
Rated power	3000 W
Rated performance	43.2 V @ 70 A
Output voltage range	39–69 V
Weight (with fan & casting)	11 kg
Reactant	Hydrogen and air
Rated H_2 consumption	42 l/min (1.48 ft^3/min)
Hydrogen pressure	0.5–0.6 Bar
Ambient temperature	5–30 °C (41–86 °F)
Max. stack temperature	65 °C (149 °F)
Hydrogen purity	99.999% dry H_2
Start-up time	immediate
Efficiency of system	40%@43.2 V

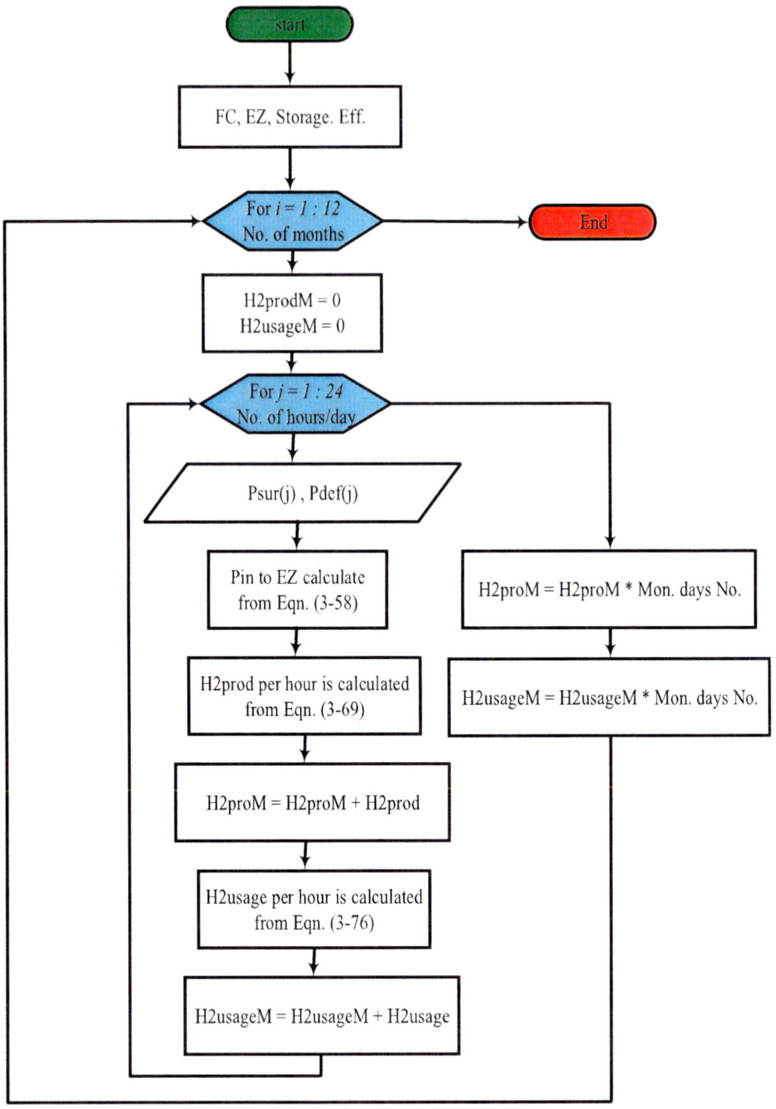

Fig. 5.24 Flowchart of a computer program for calculating hydrogen production and hydrogen usage per month

Table 5.13 Amount of hydrogen surplus/deficit hydrogen and hydrogen storage during the year in Marsa Alam

Month	Hyd. Prod. (N m^3)	Hyd. usage (N m^3)	Surp./def. (N m^3)
Jan	524.808	518.2322	6.5758
Feb	604.8388	457.6786	147.1602
Mar	719.7647	529.8409	189.9238
Apr	602.0343	521.058	80.9763
May	463.0511	563.3945	−100.3434
Jun	396.894	511.604	−114.71
July	469.5494	554.7788	−85.2294
Aug	488.4966	572.3449	−83.8483
Sep	501.1582	582.9183	−81.7601
Oct	449.9911	590.5421	−140.551
Nov	631.8178	499.1144	132.7034
Dec	559.5793	510.4332	49.1461

Table 5.14 Monthly Hydrogen production, usage and surplus/deficit in Shark El-Oweinat

Month	Hyd. Prod. (N m^3)	Hyd. usage (N m^3)	Surp./def. (N m^3)
Jan	473.7378	553.0822	−79.3444
Feb	626.9952	522.8361	104.1591
Mar	767.3553	531.9554	235.3999
Apr	598.6273	511.0751	87.5522
May	461.327	533.1806	−71.8536
Jun	440.3841	473.9994	−33.6153
July	504.2177	550.7791	−46.5614
Aug	447.8866	574.9865	−127.1
Sep	550.6347	566.5394	−15.9047
Oct	546.8841	585.9008	−39.0167
Nov	510.9127	520.1589	−9.2462
Dec	516.4195	520.5295	−4.11

5.7.4 *Final Solar Cells Area, FSCA, and the Number of Solar Panels*

Table 5.17 displays the FSCA and the number of solar panels required for supplying loads.

Applying the methodology in subitem 5.5.1 and using the buck-boost converter is a DC-DC converter for all sites to cover the weather conditions all year. Figures from 5.31, 5.32, and 5.33 display the variation of D due to the variety of input power for obtaining constant output voltage suitable for loads. At D_{min} the system operates at maximum input power, so the current output is maximum; The electrical load takes its electrical energy needs and the rest goes to storage or the utility grid. At D_{max} the

Table 5.15 Monthly Hydrogen production, usage and surplus/deficit in Minia

Month	Hyd. Prod. (N m^3)	Hyd. usage (N m^3)	Surp./def. (N m^3)
Jan	220.7867	548.5848	−327.798
Feb	590.2257	481.176	109.0497
Mar	775.0969	556.1179	218.979
Apr	719.0542	560.25	158.8042
May	493.9238	615.5438	−121.62
Jun	507.4084	595.6875	−88.2791
July	524.5117	615.5438	−91.0321
Aug	590.9723	615.5438	−24.5715
Sep	639.9506	597.8944	42.0562
Oct	649.404	599.0757	50.3283
Nov	554.5798	538.0847	16.4951
Dec	583.2203	525.5661	57.6542

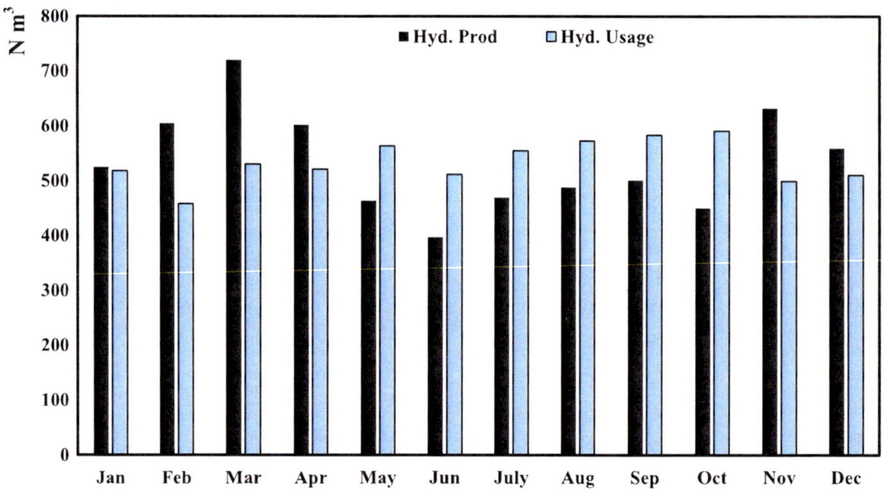

Fig. 5.25 Hydrogen production/usage (N m^3) each month during the year in Marsa Alam

system operates at minimum power that can barely or just sufficiently fulfill loads power requirement.

From the results of this chapter and previous dissection, the design of the PV power system with maximum power transfer can be shown in Table 5.18. The inverter converts 54 to 48.6 V then the transformer is a step-up voltage from 68 V to utility grid voltage (120 or 220 V) in case of connecting to a single phase.

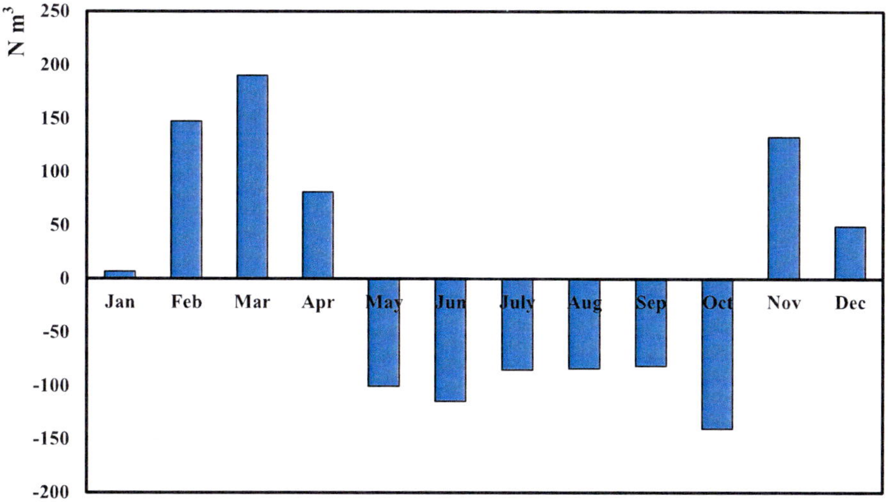

Fig. 5.26 Hydrogen surplus/deficit each month during the year in Marsa Alam

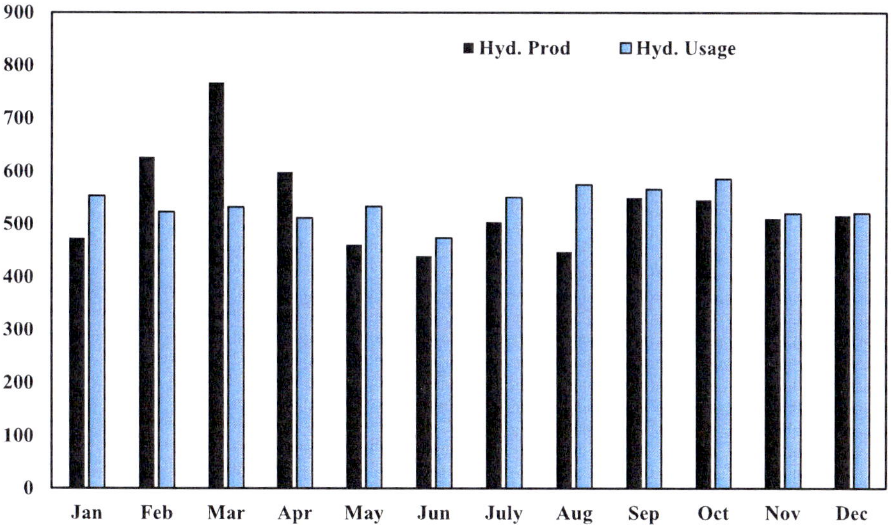

Fig. 5.27 Hydrogen production/usage (N m^3) each month during the year in Shark El-Oweinat

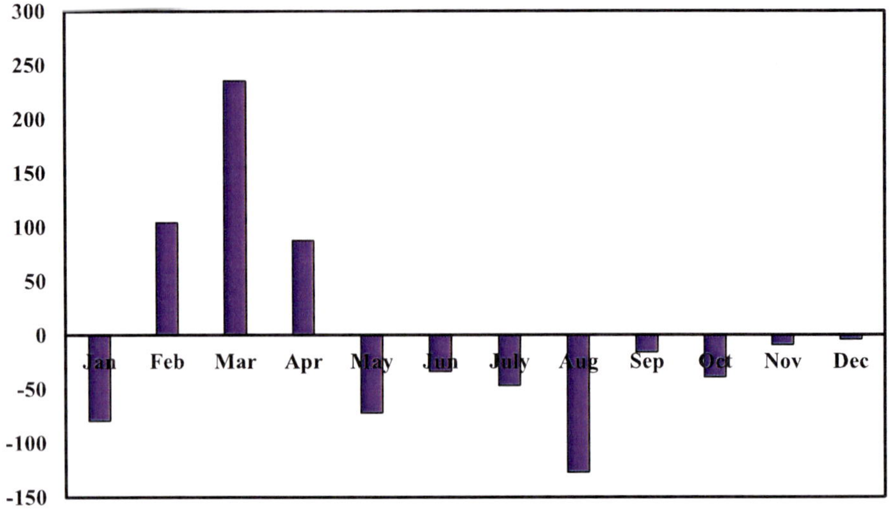

Fig. 5.28 Hydrogen surplus/deficit each month during the year in Shark El-Oweinat

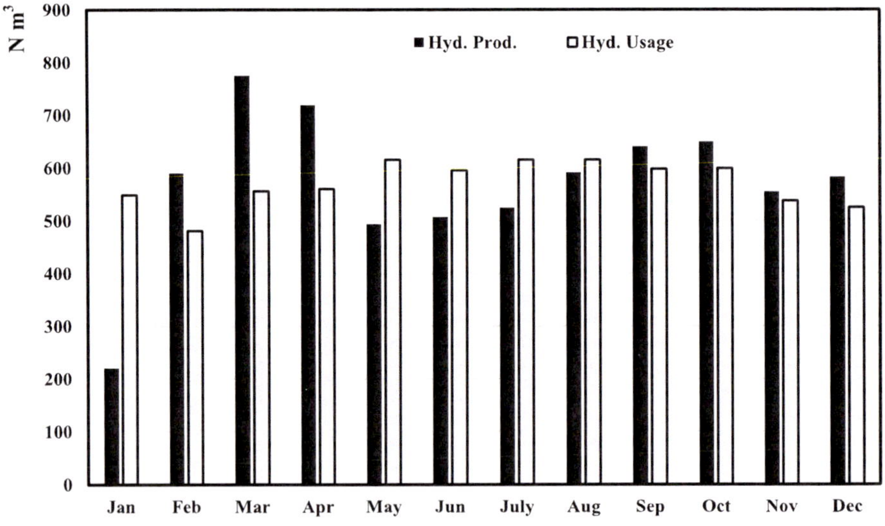

Fig. 5.29 Hydrogen production/usage (N m^3) each month during the year in Minia

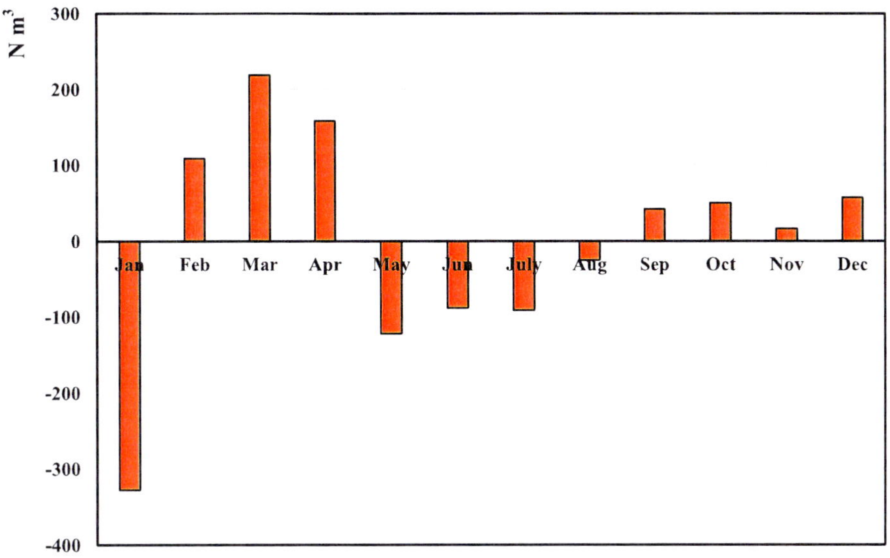

Fig. 5.30 Hydrogen surplus/deficit each month during the year in Minia

Table 5.16 Water tanks volume, m^3; and hydrogen tanks volume, N m^3 at different pressure

Site	Water tank volume, m^3	Hydrogen tank volume, at 1 bar, N m^3	Hydrogen tank volume, at 15 bar, N m^3	Hydrogen tank volume, at 30 bar, N m^3
Marsa Alam	0.6064856	606.4856	40.43237	20.21619
Shark El-Oweinat	0.4271112	427.1112	28.4741	14.237
Minia	0.4868329	486.8329	32.45553	16.22776

Table 5.17 Final SCA and number of solar panels required for feeding loads

Location	Solar module tilt angle	SCA, m^2	No. of solar panels
Marsa Alam	Monthly best tilt angle	76.9848	48
	The latitude	82.180	51
Shark El-Oweinat	Monthly best tilt angle	78.61	49
	The latitude	84.02	52
Minia	Monthly best tilt angle	84.9	53
	The latitude	90.07	56

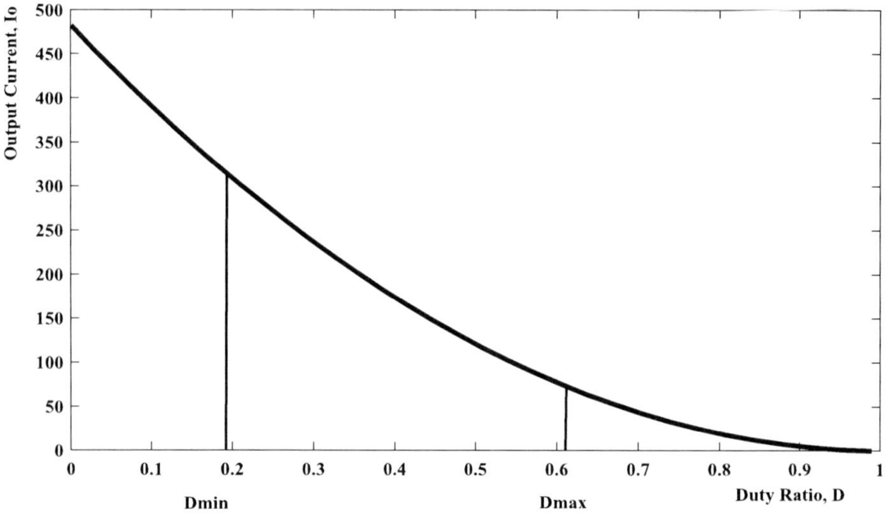

Fig. 5.31 Duty ratio vs. output current with min. and higher max. duty ratio for buck-boost converter 12 series

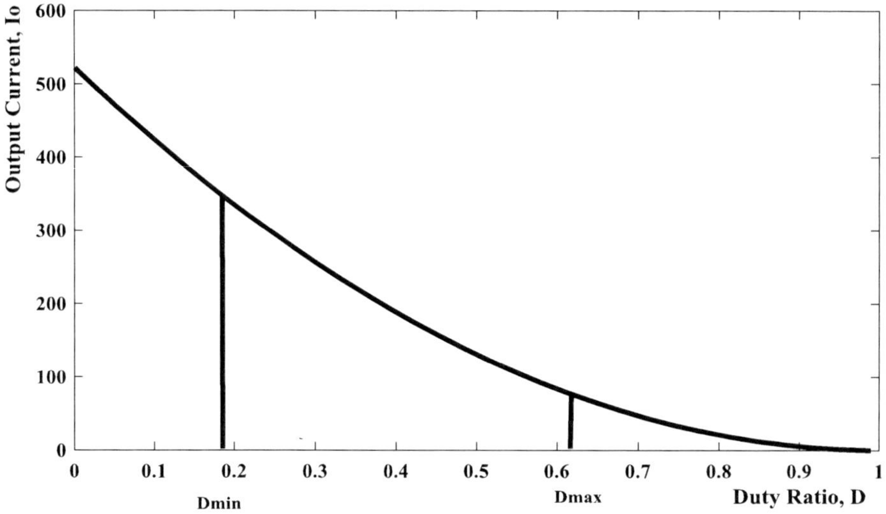

Fig. 5.32 Duty ratio vs. output current with min. and higher max. duty ratio for buck-boost converter 13 series

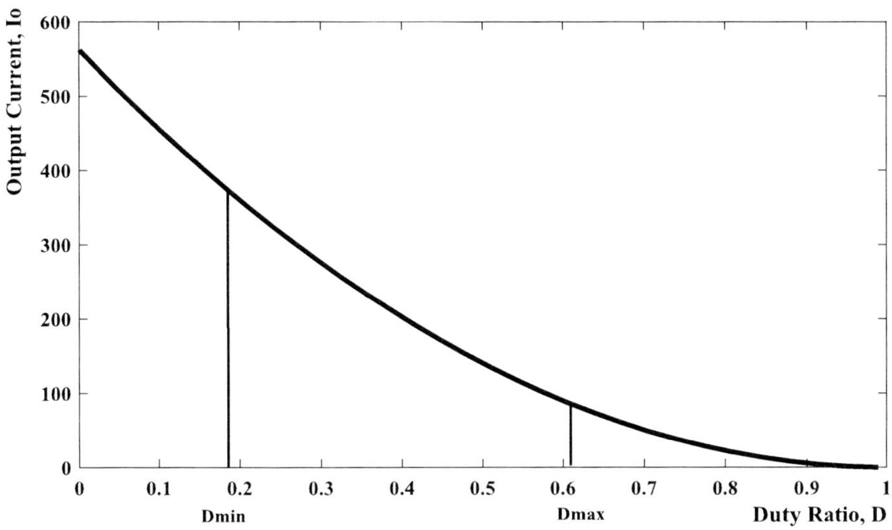

Fig. 5.33 Duty ratio vs. output current with min. and higher max. duty ratio for buck-boost converter 14 series

Table 5.18 Photovoltaic power system design details for all sites

Site	Marsa Alam		Minia		Shark El-Oweinat	
Tilt angle	Best tilt	Latitude	Best tilt	Latitude	Best tilt	Latitude
N_s	4	4	4	4	4	4
N_p	12	13	14	14	13	13
$V_{oc,a}$, (V)	278	278	278	278	278	278
$I_{sc,a}$, (A)	77.76	84.24	90.72	90.72	84.24	84.24
$V_{mp,a}$, (V)	236.4	236.4	236.4	236.4	236.4	236.4
$I_{mp,a}$, (A)	73.08	79.17	85.26	85.26	79.17	79.17
$P_{mp,a}$, (kW)	17.276	18.716	20.155	20.155	18.716	18.716
D_{min}	0.186	0.186	0.186	0.186	0.186	0.186
D_{max}	0.61	0.61	0.61	0.61	0.61	0.61
R, (ohm)	0.1688	0.1558	0.1447	0.1447	0.1558	0.1558
L, ($*10^{-6}$ H)	3.7279	3.4412	3.1954	3.1954	3.4412	3.4412
C, ($*10^{-3}$ F) at ripples 1%	7.3	8	8.6	8.6	8	8
C, ($*10^{-3}$ F) at ripples 4%	1.8	2	2.1	2.1	2	2

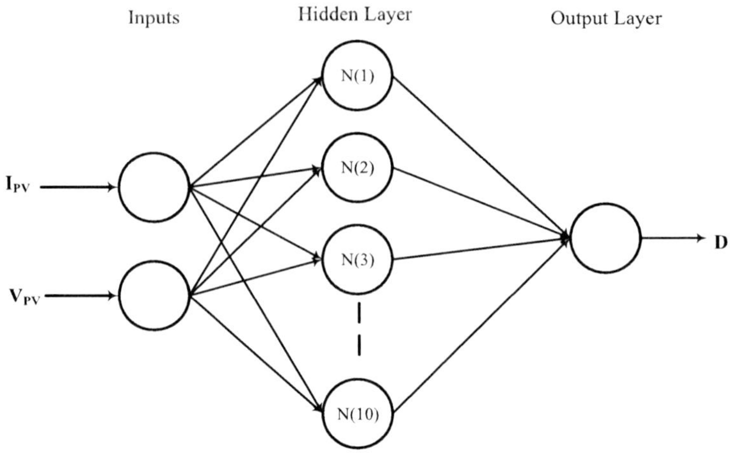

Fig. 5.34 ANN architecture (2 + 10 + 1) for determining duty cycle

Table 5.19 The weights and bases of the input layer (2 + 10 + 1) for determining duty cycle

Weights		Base
1.8052	−2.7194	−4.899
2.8374	3.4022	−3.4339
−1.5615	2.5643	2.0892
2.9887	0.70152	−1.7724
1.7697	4.4947	−1.2958
−9.27	−7.954	−2.0009
−4.6709	0.36458	−5.9592
2.3743	4.5816	2.4146
3.7325	2.3756	3.4557
3.1459	3.1191	4.4204

Table 5.20 The weights and bases of the output layer (2 + 10 + 1) ANN for determining duty cycle

Weights output									
0.83304	−1.0864	0.19597	0.51249	−1.8705	5.0953	−3.2296	−1.1906	0.63033	−0.13025
Bases output									
0.37378									

Table 5.21 The weights and bases of the input layer (2 + 10 + 1) for determining duty cycle

Weights		Bias
2.903	−2.8143	−6.3647
−0.78792	2.574	−3.0534
−2.543	3.049	0.78244
4.2083	0.086721	−1.5564
1.4809	3.0919	−0.28061
−7.514	−8.9632	−1.8975
−3.2304	0.52596	−2.5741
0.44366	4.3977	2.7391
3.5094	2.5409	3.6624
3.1563	3.1075	4.4202

5.7.5 *ANN Results for Determining the Duty Cycle*

ANN architecture (2 + 10 + 1) is designed for determining the duty cycle as shown in Fig. 5.34. The inputs are I_{PV} and V_{PV}, and the output is the value of the duty cycle for extracting maximum power (Tables 5.19, 5.20, 5.21, and 5.22). ANN technique is applied on photovoltaic power system that contain array ($N_s = 4$ modules, $N_p = 13$ branches). This string is found in Marsa Alam site and Shark El-Oweinat ANN Matlab tool is used for doing this task. Figure 5.35 displays the neural networks training tool. Validation performance is shown in Fig. 5.36. The training test results are displayed in Fig. 5.37. Figure 5.38 display the regression of ANN.

Figure 5.39 displays the outputs from the analytic method and ANN outputs. It can be noted that there are very small errors between outputs and ANN outputs.

Also, the ANN technique is applied on photovoltaic power system that contain array ($N_s = 4$ modules, $N_p = 14$ branches). This string is found in Minia. ANN Matlab tool is used for doing this task. Figure 5.40 displays the neural networks training tool. Validation performance is shown in Fig. 5.41. The training test results are displayed in Figs. 5.42 and 5.43.

Figure 5.44 displays the outputs from the analytic method and ANN outputs. It can be noted that there are very small errors between outputs and ANN outputs.

Table 5.22 The weights and bases of the output layer (2 + 10 + 1) ANN for determining duty cycle

Weights output									
−2.5309	−1.5106	0.059033	−0.92517	0.30275	5.9417	1.6943	1.324	0.043261	0.078131
Bases output									
0.60029									

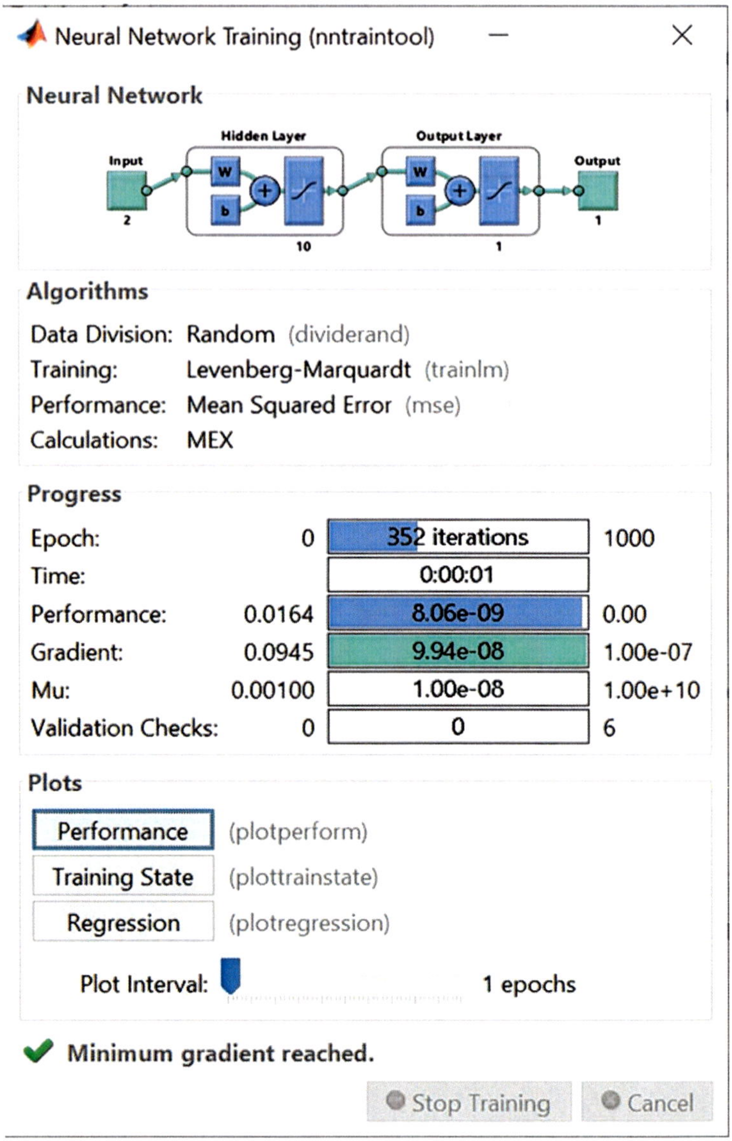

Fig. 5.35 ANN data training

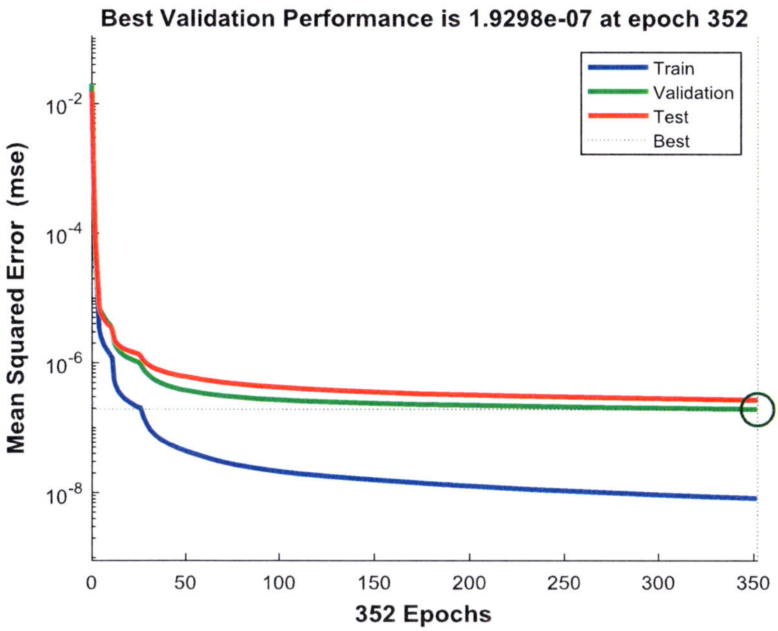

Fig. 5.36 Validation performance for ANN (2 + 10 + 1) for determining duty cycle

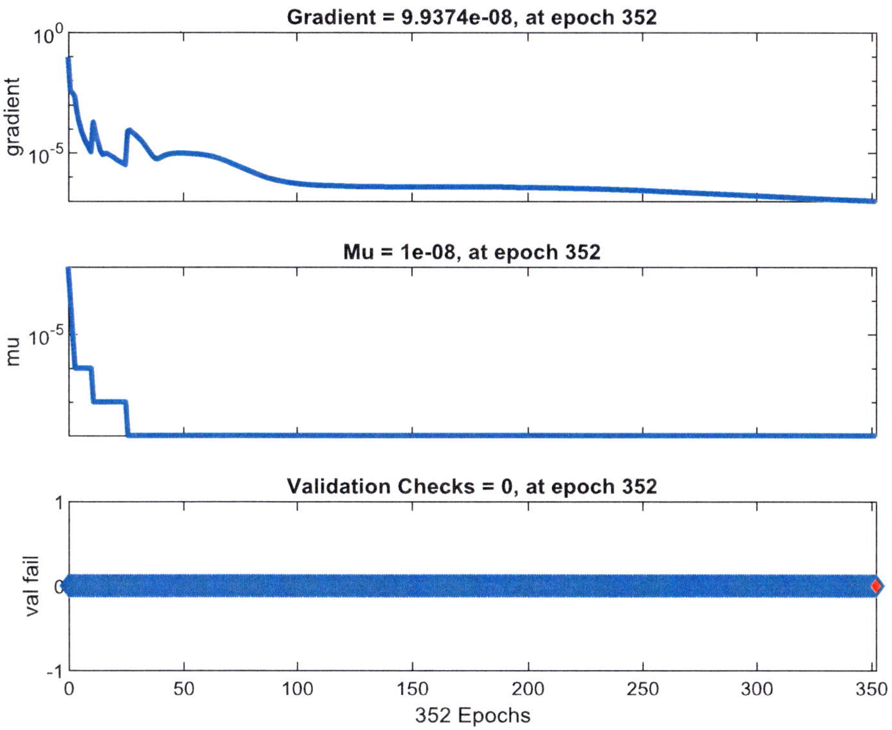

Fig. 5.37 The training test results of ANN (2 + 10 + 1) for determining duty cycle

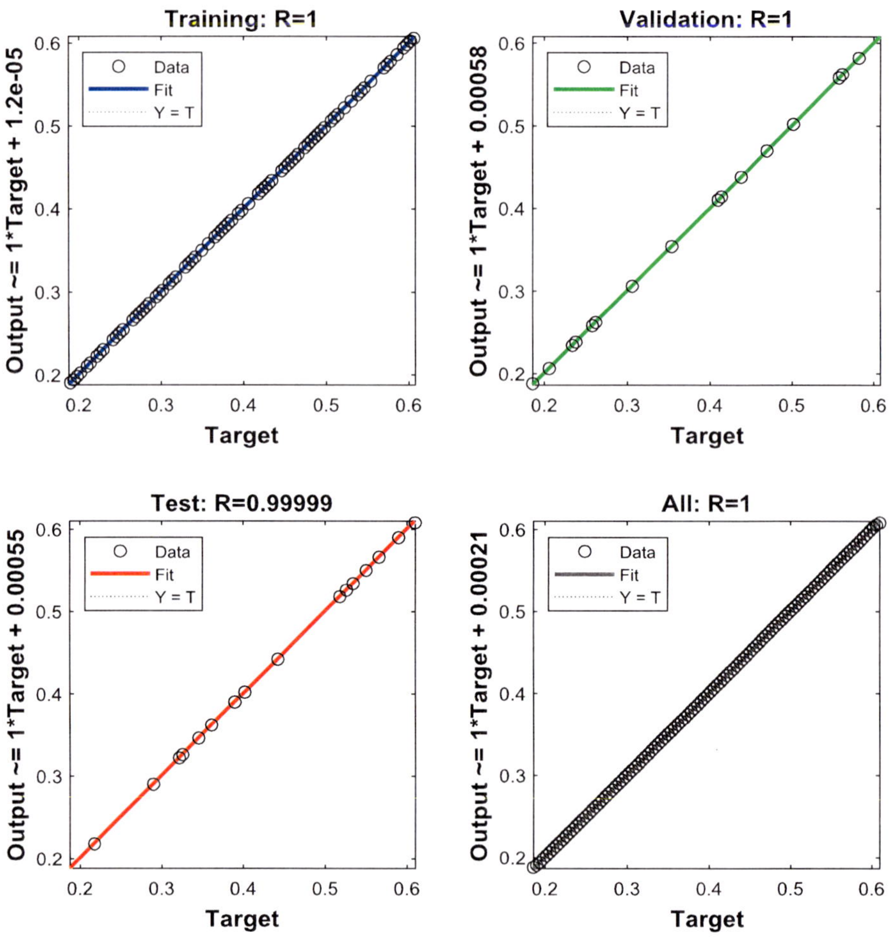

Fig. 5.38 The regression of ANN (2 + 10 + 1) for determining duty cycle

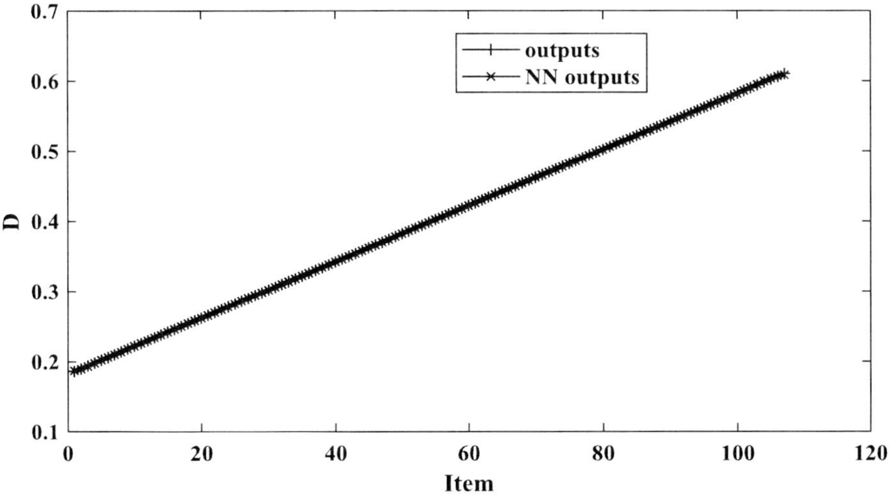

Fig. 5.39 Comparison between outputs and NN outputs

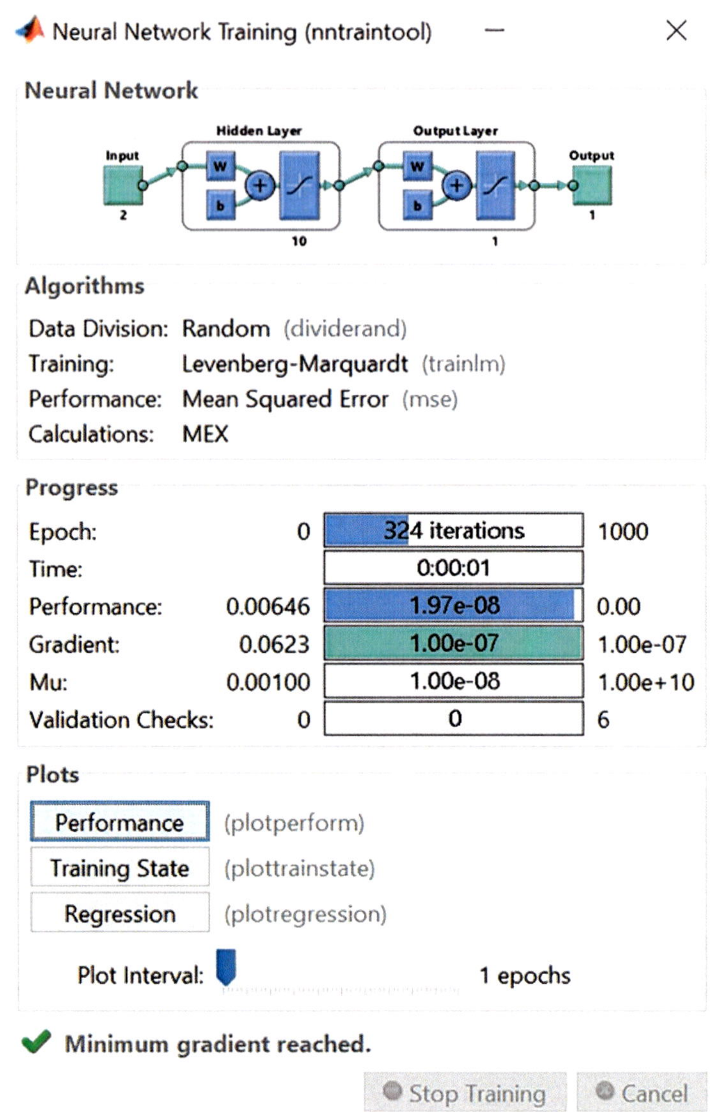

Fig. 5.40 ANN data training

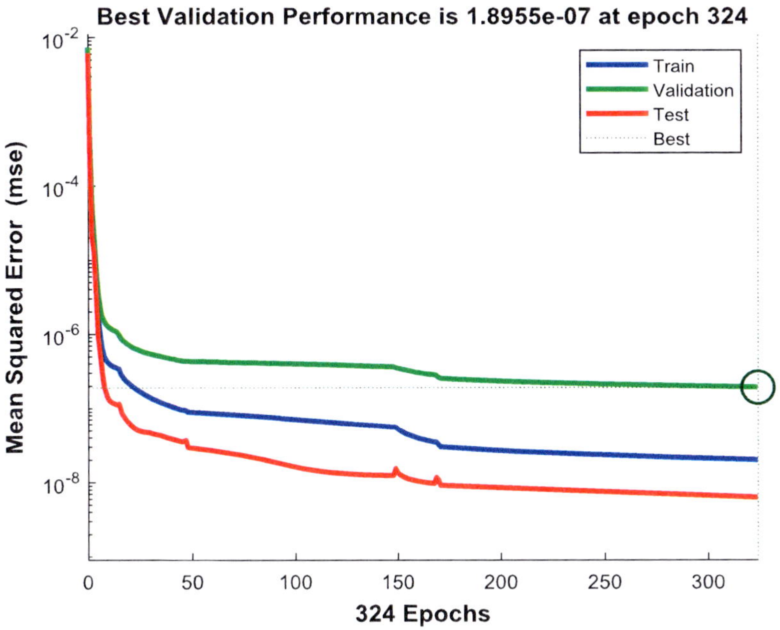

Fig. 5.41 Validation performance for ANN (2 + 10 + 1) for determining duty cycle

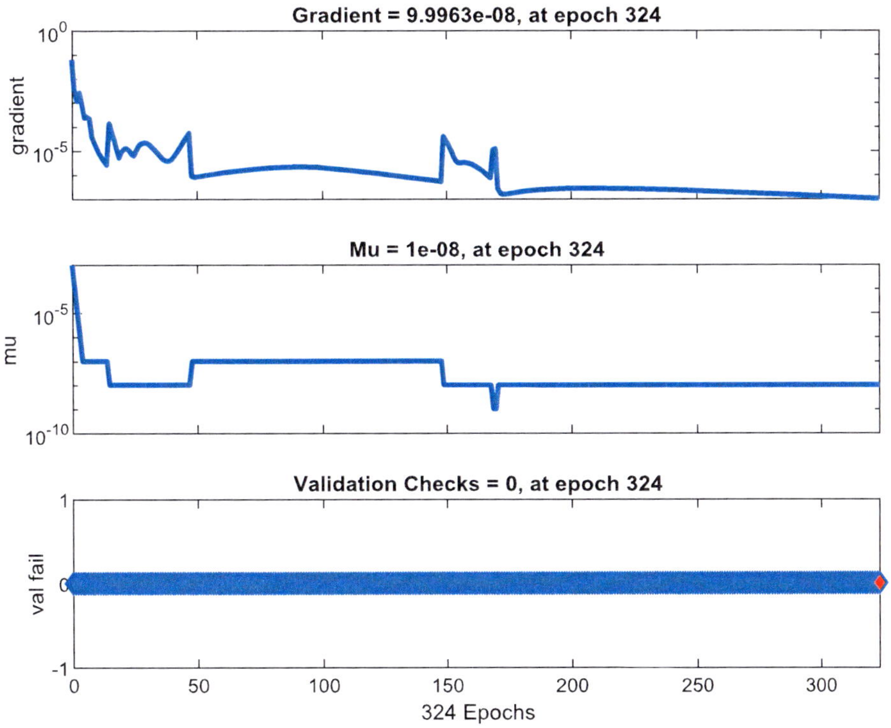

Fig. 5.42 The training test results of ANN (3 + 10 + 1) for determining duty cycle

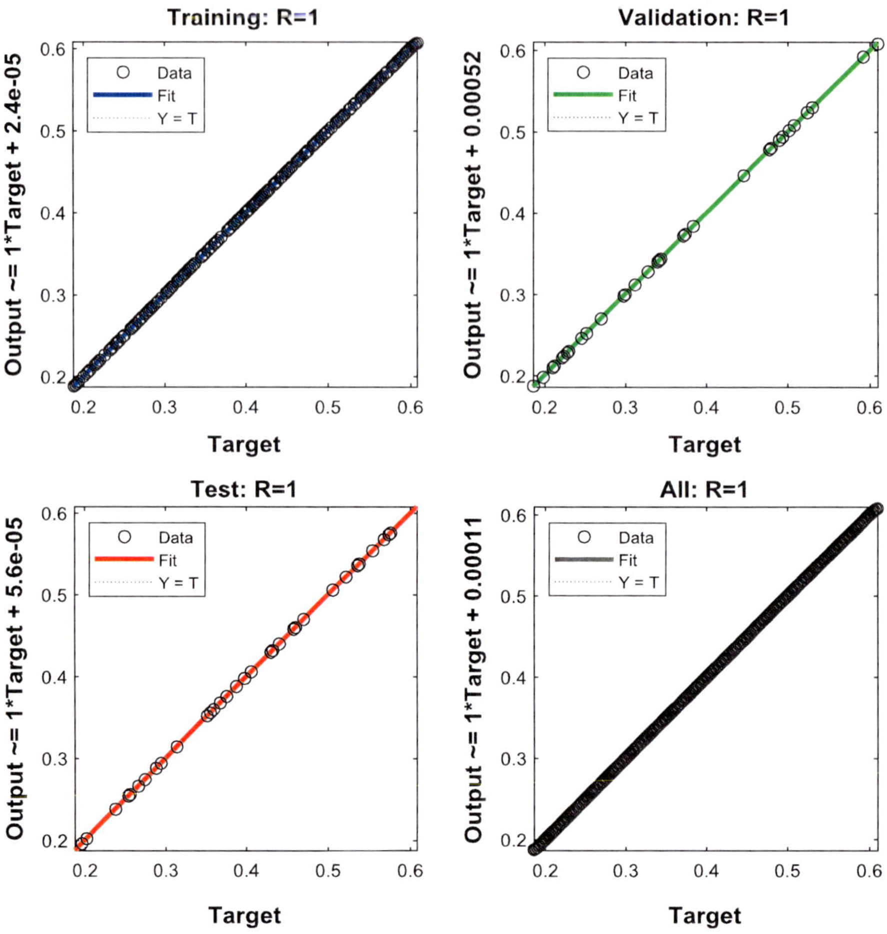

Fig. 5.43 The regression of ANN (2 + 10 + 1) for determining duty cycle

5.8 Summary

This chapter looks into the photovoltaic power system; the calculation of solar radiation incident on surfaces provided the required number of solar panels to feed a certain load. Three locations in Egypt are selected for applying the methodology. four types of modern solar panel were chosen to pick out the best of them, a computer program is prepared to calculate the number of panels required to supply

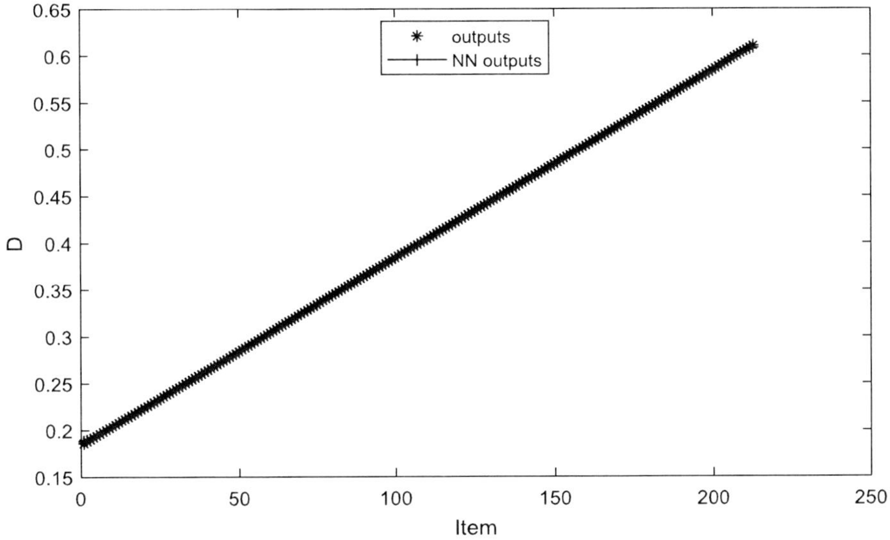

Fig. 5.44 Comparison between outputs and NN outputs

the load in cases the angle of incidence optimum monthly tilt angle as well as in the case of fixed solar panels at an angle of latitude for this site. Through what has been done, the best type of solar panels and the optimal number of each location have been identified. A study was also conducted on the connection of the photovoltaic power system to the UG. ANN technique is used to obtain MPPT.

Chapter 6
Design and Sizing Wind Energy System

6.1 Introduction

The wind is the movement of air masses originating in the atmosphere due to the difference in air pressure which in terms is generated by the heterogeneity of their heating and cooling under the influence of radiation, phase, turbulence, and heat transfer [159].

Wind energy is considered as an indirect form of solar energy as solar heating of the atmosphere along with earth's rotation creates planetary and local wind patterns. The wind arises due to uneven heating of the earth's surface. Water and territory surfaces close by clouds heat up much more slowly; accordingly, the surface of the earth accessible to sunlight heats up faster. As air passes over the warm surface, it heats up and rises, creating areas of lower pressure. Air from high pressure will move to the low-pressure areas and creating wind flow [160].

6.2 Wind Turbines

A wind turbine is a machine designed to capture the kinetic energy of the wind. It extracts some of the energy available in the wind and converts it into rotary mechanical power and in turn into electrical power. Although the configuration of wind turbines varies according to their types and designs, the main components have the same functions.

The basic components of a typical wind turbine include a rotor consisting of a number of aerodynamic shaped blades transferring the kinetic energy in the wind into rotational shaft energy, a generator which converts mechanical energy into electrical energy, a gearbox matching the speed of the rotor to that of the generator, a nacelle which protects the other parts of the wind turbine from damage, a tail vane

A. A. Elbaset, S. Ata, *Hybrid Renewable Energy Systems for Remote Telecommunication Stations*, https://doi.org/10.1007/978-3-030-66344-5_6

system which maintains the turbine pointed into the wind direction, a high-speed shaft, a low-speed shaft, and a tower on which the generator is situated [161].

According to the well-known theory of Bernoulli, this difference creates a low-pressure area above the blade, which in turn results in differential pressure between the upper and the lower surface of the blade. This produces a pressure gradient force, whose components are the lift force acting perpendicular to the airflow, and the drag force acting parallel to the direction of the airflow [162]. As a result, the rotor blades attached to a thrust shaft turn an electric generator in turbine causing it to generate electric energy.

6.2.1 Basic Components of a Wind Turbine System

Typical wind turbines involve a set of rotor blades (usually three) rotating around a hub. The hub is connected to a gearbox and a generator, located inside the nacelle, which houses the electrical components. The basic elements of a wind turbine system are shown in Fig. 6.1 and outlined as follows [163]:

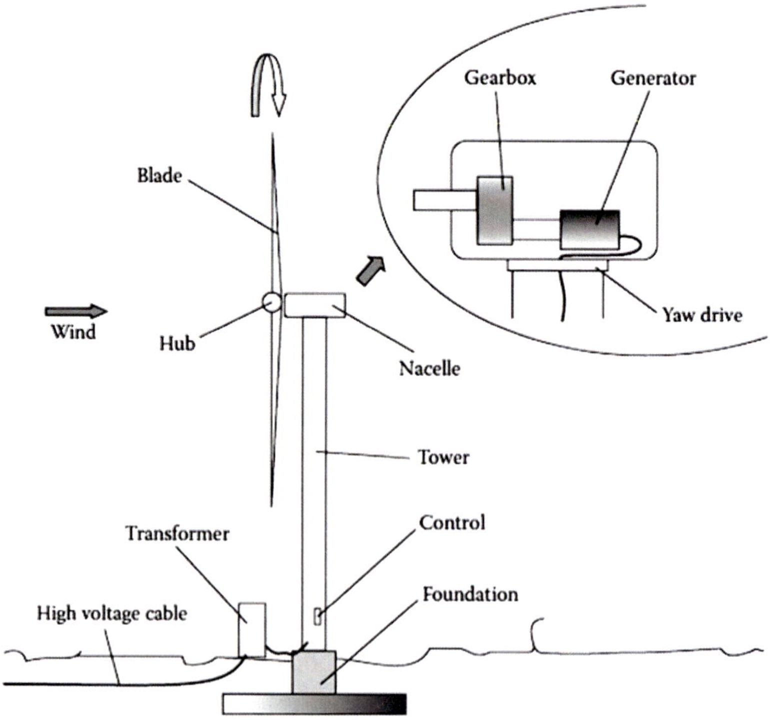

Fig. 6.1 Major turbine components

The nacelle: Sits on top of the tower and contains the electrical components—the gearbox, the brake, the wind speed and director monitor, the yaw mechanism, and the generator.

Rotor blades: The diameter of the blades is a crucial element in the turbine power; typically, the longer they are, the higher the output. But their design and the materials incorporated by them are also key elements. Blades are often made of fiberglass reinforced with polyester or wood epoxy. Vacuum resin infusion is a new material connected to a technology presented by manufacturers like Suzlon. Typically blades rotate at 10–30 revolutions per minute, either at a constant speed (the more traditional solution) or at a variable speed.

Gearboxes and direct drives: Most wind turbines use gearboxes, whose function is to increase the rotational speed required by generators. Some new technologies are exploring direct-driven generators to dispense with the expensive gears.

Brake: A disk used to stop the rotor blades in emergencies and to ensure the safety of the turbine in case of very high damaging winds or other exceptional situations.

Controller: A set of electrical components that control the starting, the stopping, and the turbine rotor blade speed. Typically, in the constant wind speed model, the controller starts up the turbine at wind speeds around 8–14 miles per hour and stops the machine at around 55 miles per hour (to avoid the damage caused by turbulent high winds).

Generator: The device responsible for the production of 60-cycle alternating current (AC) electricity.

The yaw mechanism of wind power generators: In more typical wind turbines, the yaw mechanism is connected to sensors (e.g., anemometers) that monitor wind direction, turning the tower head and lining up the blades with the wind.

Tower: Supports the nacelle and rotor. The electricity produced by the generator comes down cables inside the tower and passes through a transformer into the electricity network.

Base: Large turbines are built on a concrete base foundation. When a wind turbine ceases production, it is a simple task to dig these out or cover them, leaving a little trace behind.

6.2.2 Modeling of Wind Turbines

Power Output from an Ideal Turbine

Wind generation, like solar, tidal, and wave generation, exhibits "variable" output. The output of these units depends upon weather conditions that cannot be controlled by the operator of the generator. The use of wind for electric power generation has been increasingly considered as a possible alternative to conventional generation due to rising interest in renewable and green energy. The wind is an environmentally

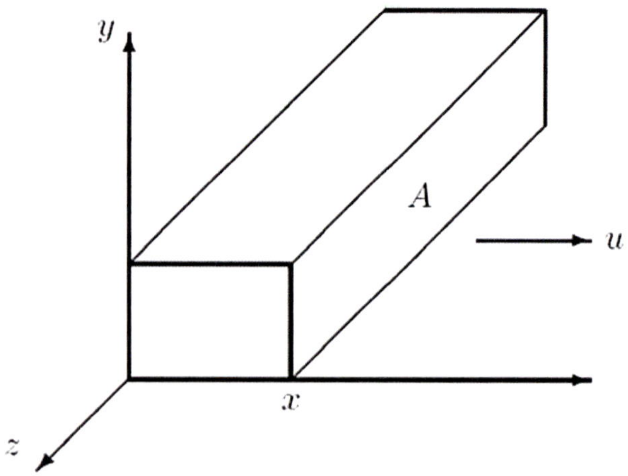

Fig. 6.2 Packet of air moving with speed u

friendly source of energy. Also, uncertain global economic and political conditions are driving many countries to be more dependent on their own natural resources and hence rely less on imported fuels. One of the main characteristics of wind is that it is highly variable and its properties vary from one location to another. The variation of wind spans from instantaneous, minute by minute, hourly, daily, seasonal, and annual [164].

A wind turbine converts the kinetic energy of the wind into rotary mechanical energy which then drives a generator. The amount of energy which the wind transfers to the rotor depends on the density of air, the rotor swept area, and the wind speed [164, 165]. The kinetic energy of the wind is defined as:

$$E = \frac{1}{2} m u^2 \tag{6.1}$$

Consider a smooth and laminar flow of wind passing perpendicularly (normally) through an element of area A of any shape, having thickness x, shown in Fig. 6.2. The mass m of air contained in an element of volume Ax is given, in terms of air density ρ, [166] by:

$$m = \rho A x \tag{6.2}$$

Combining Eqs. (6.1) and (6.2) gives, for the energy associated with this mass and volume of air:

$$E = \frac{1}{2} \rho A x u^2 \tag{6.3}$$

The relation between power and density is:

$$P_w = \frac{dE}{dt}$$
$$= \frac{d}{dt}\left(\frac{1}{2}\rho A x u^2\right)$$
$$= \frac{1}{2}\rho A u^2 \frac{dx}{dt}$$

$$\text{Then} \quad P_w = \frac{1}{2}\rho A u^3 \tag{6.4}$$

Equation (6.4) shows the greater dependency of wind power on the wind speed. An increase in wind speed by a factor of 2.1 will result in a power increase of 10. However, the effective power of wind energy conversion technology is different from the actual wind power.

Theoretical Maximum Power Extractable from the Wind

Only a fraction of the total theoretical power available in the wind, represented by Eq. (6.4), is extractable. It is an intrinsic property of all physical systems that when energy is converted from one form to another, this conversion is accompanied by various energy losses. The result is that conversion is always subject to significant intrinsic limitations of efficiency.

Let a flow of smooth and steady air with an upstream average velocity u_1 impinge upon the rotor of a wind machine [167], as illustrated in Fig. 6.3. Some of the energy from the wind is transferred to the wind machine rotor so that the smooth and steady air far downstream flows at a smaller average velocity u_2. The kinetic energy reduction of the airflow, of mass m, per unit time, is:

$$\begin{aligned} E &= \frac{1}{2}mu_1^2 - \frac{1}{2}mu_2^2 \\ &= \frac{1}{2}m\left(u_1^2 - u_2^2\right) \end{aligned} \tag{6.5}$$

In the process of extracting energy from the wind, the wind velocity u_r that actuates the rotor is less than the upstream "free wind" velocity u_1. With an ideal and

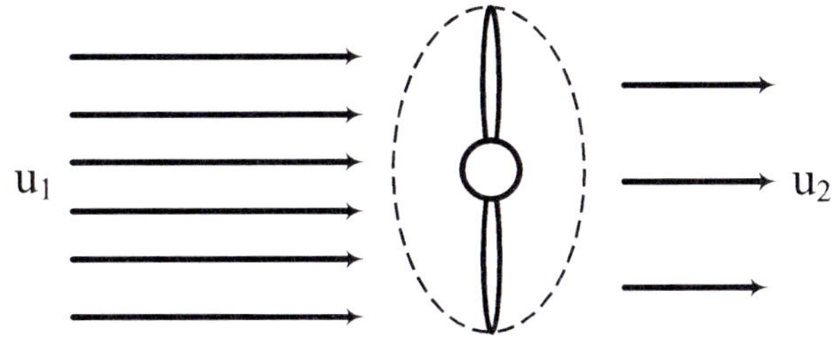

Fig. 6.3 Rotor of a wind converter

lossless system, all of the energy reduction in the airstream is transferred to the rotor of the wind machine. The downstream average velocity u_2 is then smaller than the actuating velocity u_r at the rotor. The power at the rotor is given as:

$$\begin{aligned} P_r &= \frac{1}{2}\frac{dm}{dt}\left(u_1^2 - u_2^2\right) \\ &= \frac{1}{2}\rho A\frac{dx}{dt}\left(u_1^2 - u_2^2\right) \\ &= \frac{1}{2}\rho A u_r\left(u_1^2 - u_2^2\right) \end{aligned} \tag{6.6}$$

It is found that the wind velocity u_r at the rotor may be expressed as:

$$u_r = \frac{u_1 + u_2}{2} \tag{6.7}$$

In this idealized model of airflow, the wind velocity at the rotor is, therefore, the average of the upstream and downstream steady wind velocities. Substituting for u_r from Eq. (4.12) into Eq. (4.11) gives an expression for the power extractable from the wind by the rotor. Using the symbol P_{ex} to denote extracted power,

$$\begin{aligned} P_{ex} &= \frac{1}{4}\rho A(u_1 - u_2)\left(u_1^2 - u_2^2\right) \\ &= \frac{1}{4}\rho A u_1^3\left[1 + \frac{u_2}{u_1} - \left(\frac{u_2}{u_1}\right)^2 - \left(\frac{u_2}{u_1}\right)^3\right] \\ &= \frac{1}{4}\rho A u_1^3\left[1 + \frac{u_2}{u_1}\right]\left[1 - \left(\frac{u_2}{u_1}\right)^2\right] \end{aligned} \tag{6.8}$$

The value of wind velocity ratio (u_2/u_1) that results in maximum power transfer is calculated by differentiating equation (6.8) with respect to (u_2/u_1) and equating to zero. This results in a quadratic equation showing that to maximize P_{ex} the ratio (u_2/u_1) must have the values either $(u_2/u_1) = 1/3$ or $(u_2/u_1) = -1$.

The negative option is meaningless so that the correct solution is:

$$\frac{u_2}{u_1} = \frac{1}{3} \tag{6.9}$$

Substituting Eq. (6.8) into Eq. (6.9) gives an expression for the maximum possible power extraction, under ideal conditions.

$$\begin{aligned} P_{ex,\max} &= \frac{8}{27}\rho A u_1^3 \\ &= \left(\frac{16}{27}\right)\frac{1}{2}\rho A u_1^3 \\ &= (0.593)\frac{1}{2}\rho A u_1^3 \end{aligned} \tag{6.10}$$

The very important result (6.10) is sometimes referred to as Betz Law. This states that "even with ideal wind energy conversion the maximum power transferable is only 0.593 or 16/27 of the total power in the wind." In reality, only a fraction of this theoretical maximum power is realized, even by the best-designed turbines [168].

Practical Power Extractable from the Wind and Power Coefficient

The power actually extractable by a wind turbine is much less than the maximum theoretical value defined in Eq. (6.10). A practical wind machine experiences air drag and air friction on the rotor blades causing heat losses. In addition, the rotation of the blades causes swirling of the air and eddies, which reduce the torque imparted to the blades. The net effect of the various losses is incorporated into a parameter called the power coefficient C_p. With an upstream velocity u_1, the extractable power P_{ex} can be written as:

$$P_{ex} = C_p \frac{1}{2}\rho A u_1^3 \tag{6.11}$$

Coefficient C_p is seen to be the ratio of the power P_w in the wind (6.3) and the power P_{ex} extracted (6.8). It, therefore, represents the efficiency of the turbine rotor:

$$C_p = \frac{P_{ex}}{P_w} \tag{6.12}$$

Parameter C_p is a dimensionless variable. By comparison of Eq. (6.10) with Eq. (6.7), it is seen that C_p can be expressed in terms of the upstream and downstream average wind speeds.

$$C_p = \frac{1}{2}\left[1 + \frac{u_2}{u_1}\right]\left[1 - \left(\frac{u_2}{u_1}\right)^2\right] \tag{6.13}$$

For the ideal theoretical case, when $(u_2/u_1) = 1/3$, coefficient C_p has a maximum or Betz Law value of 0.593. But for practical wind turbines, the value is usually in the range $0 \le C_p \le 0.4$. With an amount $C_p = 0.4$, for example, the power available from the wind is (0.4/0.593) or about 67% of the ideal theoretical value and is 40% of the total power in the wind. Power coefficient C_p has an amount that depends on the average wind velocity, the turbine rotational velocity, and also on turbine blade design parameters such as the pitch angle [169].

Efficiency Considerations of Wind-Powered Electricity Generation

The overall efficiency η can be represented in terms of the individual stage efficiencies as:

$$\eta = C_{\mathrm{p}} \cdot \eta_{\mathrm{gb}} \cdot \eta_{\mathrm{g}} \tag{6.14}$$

For small wind energy installation, up to a few kW output rating, the overall efficiency η is of the order 20–25% [122]. The electrical output power may be written in terms of the total power in the wind P_{w} and the wind velocity V utilizing Eq. (6.14) and substituting from the basic Eq. (6.3).

$$P_{\mathrm{e}} = C_{\mathrm{p}} \cdot \eta_{\mathrm{gb}} \cdot \eta_{\mathrm{g}} \cdot \frac{1}{2}\rho A u_1^3 \tag{6.15}$$

The value of the power coefficient C_{p} for the two-blade propeller type of the wind turbine varies with size and rating. They are found to be in the ranges shown below.

$$\begin{aligned} C_{\mathrm{p}} &= 0.4 - 0.5 \text{ for large machines}(100 \text{ kW} - 3 \text{ MW}) \\ &= 0.2 - 0.4 \text{ for small machines}(1 \text{ kW} - 100 \text{ kW}) \\ &= 0.35 \text{ typically for small machines} \end{aligned} \tag{6.16}$$

Modern types of mechanical gearbox have efficiencies that depend both on the size (rating) and the speed of rotation. At the rated speeds, it is found that the gearbox efficiencies lie in the following ranges.

$$\begin{aligned} \eta_{\mathrm{gb}} &= 80 - 95\% \text{ for large machines and} \\ &= 70 - 80\% \text{ for small machines} \end{aligned} \tag{6.17}$$

Electrical generators have efficiencies that increase with rated size. The process of converting rotational mechanical energy to electrical energy is inherently more efficient than wind turbine conversion or any process involving heat to work (i.e., thermodynamic) conversion. For operation at the rated output:

$$\begin{aligned} \eta_{\mathrm{g}} &= 80 - 95\% \text{ for large machines} \\ &= 60 - 80\% \text{ for small machines} \end{aligned} \tag{6.18}$$

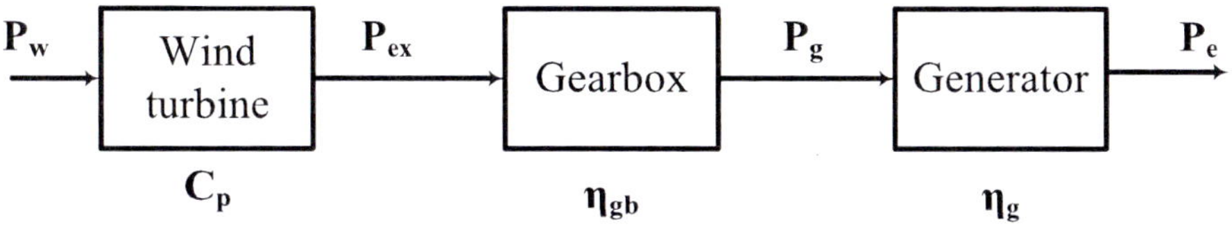

Fig. 6.4 Power flow in a wind converter system

Notably, large systems are more efficient than small systems. In general, an increase in scale is accompanied by an increase in efficiency [122]. Typical values of overall efficiency η can be obtained by substituting typical values of stage efficiencies into Eq. (6.17) (Fig. 6.4).

$$\begin{aligned} &\text{For large machines}: \quad \eta = 0.45 \times 0.9 \times 0.9 = 0.36 \\ &\text{For small machines}: \quad \eta = 0.35 \times 0.75 \times 0.7 = 0.18 \end{aligned} \tag{6.19}$$

6.3 Wind Parameter

6.3.1 Wind Speed

The major characteristics defining the power value of wind is its speed. Along with wind speed, the wind direction also plays a vital role in power generation. The wind speed and direction vary with locations and seasons. Generally in winter wind speed is higher than summer. Diurnal variation of wind speed is also observed as a rule, near to the seas and big lakes. In the morning wind is blowing in a coast direction, but as the sun heats the earth faster than water the wind shifts and in the evening it blows in the opposite directions as the earth cools down faster than water [170].

6.3.2 Wind Classes

Wind classes are related to a classification scheme used in the wind industry that is based on annual average wind speed. The wind strength at a particular location is classified based on power density at an elevation of 50 m above ground level. Wind power density along with wind speed for different wind classes is summarized in Table 6.1 [171].

Table 6.1 Wind classes

Wind class	Wind speed (m/s)	Wind power density (W/m^2)
1	<5.6	<200
2	5.6–6.4	200–300
3	6.4–7.0	300–400
4	7.0–7.5	400–500
5	7.5–8.0	500–600
6	8.0–8.8	600–800
7	>8.8	800–2000

6.3.3 Wind Shear

Wind shear describes the change in wind with different heights. To define wind shear, two different methods are followed: power-law profile and logarithm profile. Among this power-law profile is mostly used to determine the wind speed at a different altitude. In a power-law profile, the relation between wind speed and height is summarized by the following equation [122]:

$$\frac{u}{u_o} = \left(\frac{h}{h_o}\right)^{\gamma} \tag{6.20}$$

where u and u_o are wind speeds at heights h and h_o and γ is the wind shear constant. The value of the wind shear constant depends on the height and surface roughness.

Another method to determine the wind speed at various heights is the logarithm profile which is described by the Eq. (6.21) where the roughness of the surface is represented in [171, 172].

$$\frac{u}{u_o} = \frac{\ln(h/z_o)}{\ln(h_o/z_o)} \tag{6.21}$$

Based on Eqs. (6.20) and (6.21), the value of wind shear constant can be derived as follows [172]:

$$\gamma = \frac{\ln(\ln(h/z_o)/\ln(h_o/z_o))}{\ln(h/h_o)} \tag{6.22}$$

From Eq. (6.22) it is noted that wind shear is a function of heights and roughness length.

Wind speed does not increase with height indefinitely, even at a slower rate. The data collected at Merida airport in Mexico show that typically wind speed increases with height up to about 450 m and then decreases (Fig. 6.6). The wind speed at 450-m height can be 4–5 times greater than that near the ground surface. Modern wind turbines operate on increasingly taller towers to take advantage of the increased wind speeds at higher altitudes.

6.3.4 Weibull Distribution

This is the general model that describes the wind speed variations. This model optimizes the turbine design to minimize electricity production costs. The Weibull coefficient reflects the distribution of wind speeds, and it is determined by the Weibull distribution curve [173]. It is very important for the wind industry to be able to relatively simply describe the wind regime on-site. Turbine designers need the information to optimize the design of their turbines, to minimize generating costs. Turbine investors need the information to estimate their income from electricity generation.

The bin width is typically 1 m/s. The histogram provides information on how often the wind is blowing for each wind speed bin. The histogram for a typical site can be presented using the Weibull distribution expressing the frequency distribution of the wind speed in a compact form. The two-parameter Weibull distribution is described mathematically as [171] (Fig. 6.5):

$$f(u) = \left(\frac{K}{C}\right)\left(\frac{u}{C}\right)^{K-1} \exp\left(-\left(\frac{u}{C}\right)^{K}\right) \quad (K > 1, u > 0, C > 1) \tag{6.23}$$

where $f(u)$ is the frequency of occurrence of wind speed u. The scaling factor C is a measure for the wind speed while the shape factor k describes the shape of the

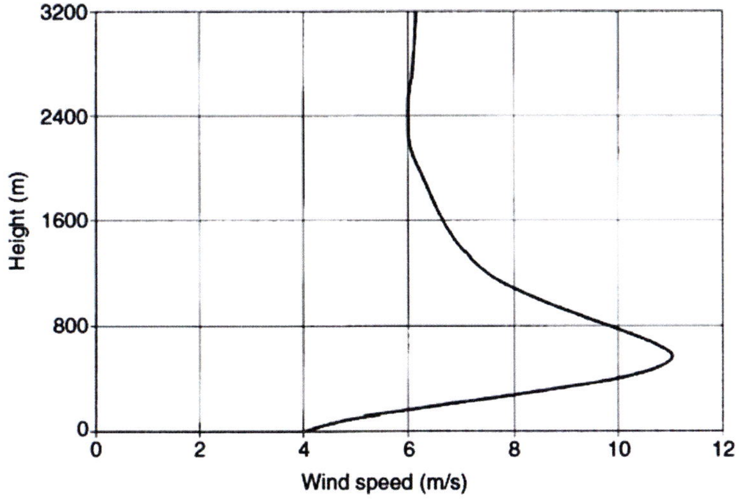

Fig. 6.5 Wind speed variations with height

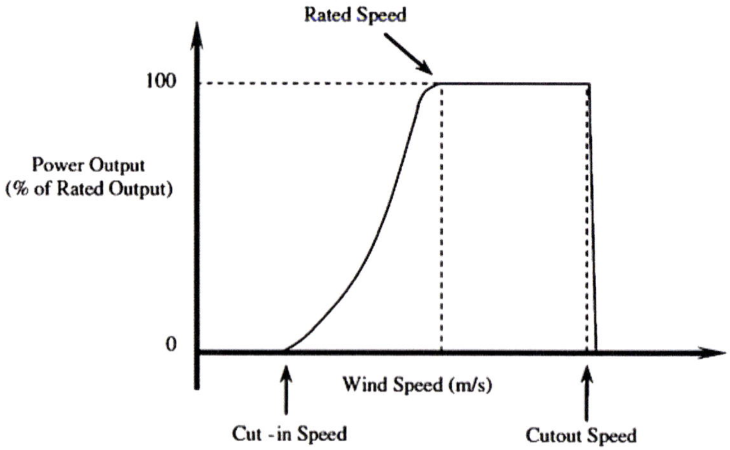

Fig. 6.6 Wind turbine power output variation with wind speed

distribution. The cumulative Weibull distribution $F(u)$ gives the probability of the wind speed exceeding the value u and is given by the simple expression [174]:

$$F(u) = 1 - \exp\left(-\left(\frac{u}{C}\right)^K\right) \tag{6.24}$$

Weibull parameters C and K can be estimated by Matlab functions depending on wind speeds distribution. There are many methods to estimate Weibull parameters that can be found in [175].

6.3.5 Wind Turbine Power Output

As discussed, the kinetic energy of the wind is first converted into mechanical energy and then into the electrical energy of the generator. Figure 6.6 illustrates the power curve of a wind turbine which describes the relationship between power and wind speed. The power curve of a wind turbine is a graph representing the relationship between the power output of the turbine and wind speed. In other words, it illustrates the quantity of power output from the wind turbine over a wide range of wind speeds. The power curve differs significantly according to the design and size of the wind turbine. The figure below shows a sketch of how the power output from a wind turbine varies with steady wind speed [176].

Cut-in Wind Speed, u_c

At very low wind speeds, there is insufficient torque exerted by the wind on the turbine blades to make them rotate. However, as the speed increases, the wind turbine will begin to rotate and generate electrical power. The speed at which the turbine first starts to rotate and generate power is called the ***cut-in speed*** and is typically between 3 and 4 m/s.

Rated Wind Speed, u_r

As the wind speed rises above the cut-in speed, the level of electrical output power rises rapidly as shown. However, typically somewhere between 12 and 17 m/s, the power output reaches the limit that the electrical generator is capable of. This limit to the generator output is called the ***rated power output*** and the wind speed at which it is reached is called the ***rated output wind speed.*** At higher wind speeds, the design of the turbine is arranged to limit the power to this maximum level, and there is no further rise in the output power. How this is done varies from design to design but typically with large turbines, it is done by adjusting the blade angles so as to keep the power at the constant level.

Cut-Out Wind Speed, u_f

As the speed increases above the rate output wind speed, the forces on the turbine structure continue to rise and, at some point, there is a risk of damage to the rotor. As a result, a braking system is employed to bring the rotor to a standstill. This is called the ***cut-out speed*** and is usually around 25 m/s [177].

The characteristics of the power output from the wind turbine generator, WTG, can be illustrated in Fig. 6.8. The characteristics can be described by the following equation:

$$P_e = \begin{cases} 0, & u \leq 0 \\ A + B \times u^K, & u_c < u < u_r \\ P_{er}, & u_r \leq u < u_f \\ 0, & u \geq u_f \end{cases} \tag{6.25}$$

where P_{er} is the rated electrical power, W, u_c is cut-in wind speed, m/s, u_r is the rated wind speed, m/s, u_f is furling wind speed, m/s.

A, B: coefficients.

$$A = \frac{\left(P_{\mathrm{er}} \cdot u_{\mathrm{c}}^{K}\right)}{\left(u_{\mathrm{c}}^{K} - u_{\mathrm{r}}^{K}\right)}, \quad B = \frac{P_{\mathrm{ewr}}}{\left(u_{\mathrm{r}}^{K} - u_{\mathrm{c}}^{K}\right)} \tag{6.26}$$

6.4 Average Power Output and Average Number of Wind Turbine Generators

The power output of WTG can be adequately described by the model of equation (6.25). A combination of the variation in output power with wind speeds and the variation in the wind speed at a site must be done to have the average power $P_{\mathrm{e,av}}$, that would be expected from a given WTG at a given location. The average power output from a WTG is a very important parameter of a wind energy system since it determines the total energy production and the total income. It is a much better indicator of economics than the rated power [178].

The average power output from the wind turbine generator is:

$$P_{\mathrm{e,av}} = P_{\mathrm{er}} \left\{ \frac{\exp\left[-(u_{\mathrm{c}}/C)^{K}\right] - \exp\left[-(u_{\mathrm{r}}/C)^{K}\right]}{(u_{\mathrm{r}}/C)^{K} - (u_{\mathrm{c}}/C)^{K}} - \exp\left[-\left(\frac{u_{\mathrm{f}}}{C}\right)^{K}\right] \right\} \tag{6.27}$$

From Eq. (6.26) $P_{\mathrm{e,av}}$ is affected by cut-in, rated and furling wind speeds for a given wind regime. A WTG must be selected to maximize the average power, and thereby maximize the total energy production.

The quantity inside the brackets in Eq. (6.27) is called the capacity factor, CF, also called the plant factor; it is an important design item in addition to the average power.

$$\mathrm{CF} = \frac{\exp\left[-(u_{\mathrm{c}}/C)^{K}\right] - \exp\left[-(u_{\mathrm{r}}/C)^{K}\right]}{(u_{\mathrm{r}}/C)^{K} - (u_{\mathrm{c}}/C)^{K}} - \exp\left[-\left(\frac{u_{\mathrm{f}}}{C}\right)^{K}\right] \tag{6.28}$$

Then,

$$P_{\mathrm{e,av}} = P_{\mathrm{er}} \cdot (\mathrm{CF}) \tag{6.29}$$

The average number of the wind turbine generator, ANWTG, can be estimated by knowing the average power production, $P_{\mathrm{e,av}}$, and average load power, $P_{\mathrm{L,av}}$, which can be determined from Eq. (6.30)

$$P_{\mathrm{L,av}} = \frac{1}{8760} \cdot \sum_{i=1}^{8760} P_{\mathrm{Lh}}(i) \tag{6.30}$$

where P_{Lh} is the hourly load.

Then ANWTG required is given by the following equation:

$$\mathrm{ANWTG} = \frac{P_{\mathrm{L,av}}}{P_{\mathrm{e,av}}} \tag{6.31}$$

6.5 Design Algorithms

The steps of design of WES are summarized in the following subsections:

6.5.1 *Upgrading Wind Velocities to Hub Height*

Usually, weather stations measure the wind speed at 10-m or 20-m. If these heights do not match the hub height of WTG, it is necessary to extrapolate the wind speeds to the hub height of WTG. This process can be done as discussed in subitem 6.3.3.

6.5.2 *Evolution of Weibull Parameter (C and K)*

Weibull parameters can be determined by using wind speeds at hub height and Matlab Weibull functions.

6.5.3 *Evolution of the Capacity Factor, CF, and ANWTG*

The capacity factor, CF, is calculated by using Eq. (6.28) for each WTG type. The average electrical power output, $P_{\mathrm{e,av}}$, for each WTG can be calculated using Eq. (6.29). Then the average number of WTG, ANWTG, can be calculated by using Eq. (6.31) for each type of WTG in each location.

6.5.4 Energy Balance Study and the Optimum Number of WTG

An energy balance between the loads and the output power from WES must be made to obtain the optimum number of WTG, ONWTG. The output energy from WES must satisfy the load-required energy. For periods with high wind speed (i.e., the output power from WES is greater than the load required), storage equipment will take the surplus power. For the periods of low output power from WES, which is less than the load demand storage equipment will give the deficit power. The ONWTG required, and power output for each case can be obtained as follows:

1. The hourly generated power can be calculated according to the value of hourly wind speed and specs of WTG that used, from the following equation:

$$P_e(t) = \begin{cases} 0, & u(t) \leq 0 \\ \left(A + B \times u^K\right) \cdot N, & u_c < u(t) < u_r \\ P_{er} \cdot N, & u_r \leq u(t) < u_f \\ 0, & u(t) \geq u_f \end{cases} \quad (6.32)$$

   where $P_e(t)$ is the hourly generated power, kW, $u(t)$ is the hourly wind speed, m/s, and N is the number of WTGs.
2. A comparison between the hourly generated power, $P_e(t)$, and hourly load $P_{Lh}(t)$ is done every hour of the year. If $P_e(t)$ is greater than $P_{Lh}(t)$, then there is hourly surplus power, but if $P_e(t)$ is smaller than $P_{Lh}(t)$ then there is an hourly deficit power. At any value of N, if the summation of hourly surplus power equals the summation of hourly deficit power then this value of N represents the optimum number of WTGs, ONWTG. To have ONWTG the process is made:

   (a)

$$\text{If} \quad \sum_{t=1}^{t=8760} [P_e(t) - P_{Lh}(t)] > 0 \quad (6.33)$$

   Then, the number of WTG must be decreased by one unit of WTG and repeating the previous process.

   (b)

$$\text{If} \quad \sum_{t=1}^{t=8760} [P_e(t) - P_{Lh}(t)] < 0 \quad (6.34)$$

   Then, the number of WTG must be increased by one unit of WTG and repeating the previous process.

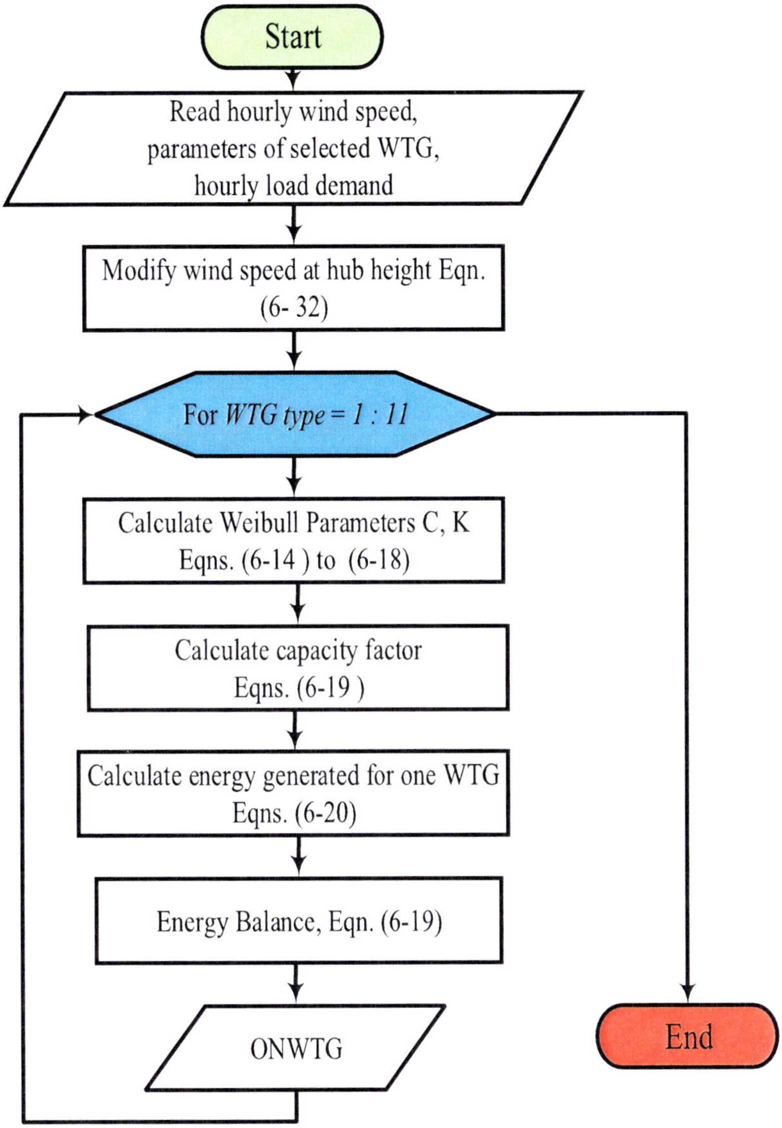

Fig. 6.7 Flow chart of a computer program to estimate ONWTG

(c)
$$\text{If} \quad \sum_{t=1}^{t=8760} [P_c(t) - P_{Lh}(t)] = 0 \tag{6.35}$$

Then, the number of WTG satisfies the energy balance condition, and then the value of N is taken as ONWTG.

A computer program is designed to estimate ONWTG. Flowchart of this computer program can be shown in Fig. 6.7.

6.6 Stand-Alone (Off-Grid) Wind Power System

The stand-alone (off grid) wind energy required to feed electrical loads in communication station contains main power system represented in wind turbine, and backup power system represented in fuel cell stack; UC is used for uninterrupted energy flow to loads; other electronic devices are used for conditioning electric power; Fig. 6.8 shows overall system.

6.7 On-Grid Wind Energy Systems

The grid-connected WECS is the overall system that converts the wind energy into useful electrical power that delivered to the electrical grid through the power transformer and transmission line. The grid-connected WECS can be represented using its principal components as shown in Fig. 6.9. It typically consists of a wind

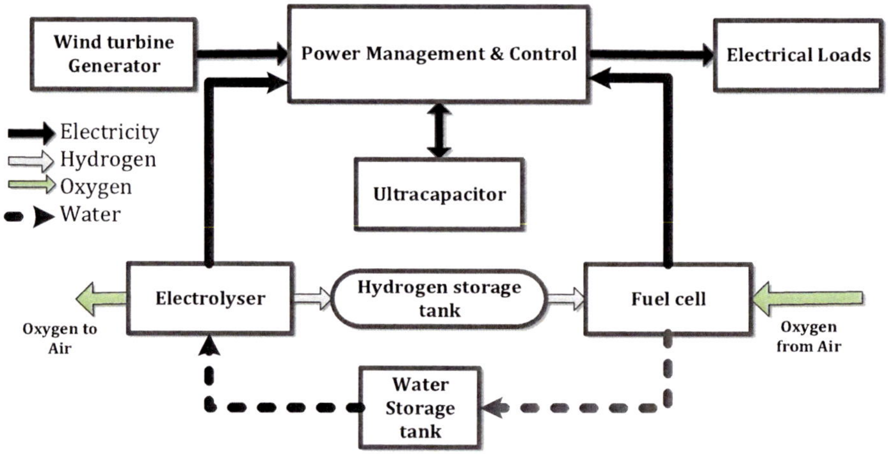

Fig. 6.8 Off-grid wind energy system for feeding communication station

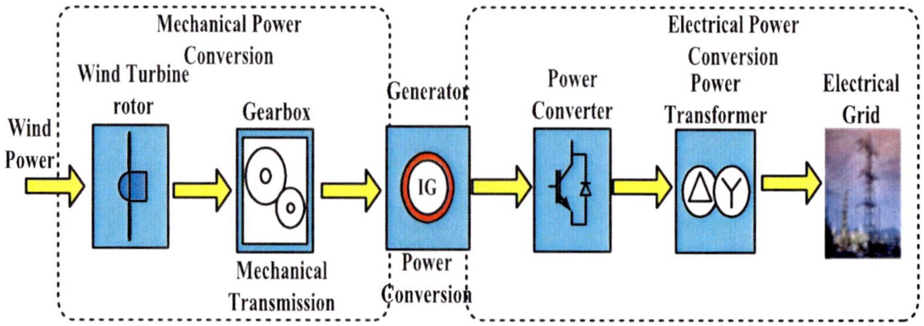

Fig. 6.9 Block diagram of the wind energy conversion system

turbine, mechanical system (drive train), electrical generator, power electronic interface, and control system components. The wind turbine converts the kinetic energy from the wind into mechanical energy that transmitted through the drive train to the electrical generator. Then, the electrical generator utilizes this mechanical energy to generate the electrical power that directly injected into the electrical grid or rural areas [179]. In the following, the principal components employed in the grid-connected wind energy conversion system are discussed in detail.

A small-scale wind turbine consists of a generator, a power electronic converter, and a control system. Among different types of a small-size wind turbine, a permanent magnet (PM) generator is widely used because of its high reliability and simple structure [180]. The power electronic converter topology used depends on the required output power and cost of the system. Control systems are used to control the rotational speed of small-scale wind turbines enabling them to operate with optimum speed to extract maximum power from wind [181].

There is an increasing market for a grid-connected small wind generating system for homeowners and small businesses in rural areas. In this case, the excess energy from the wind generator is fed to the utility grid. A grid-connected inverter structure that extracts energy even at low wind speeds will assist in reducing capital cost and offer opportunities for interfacing small-scale wind generators with the AC grid. Conventional grid-connected wind turbines use a charge controller to bypass energy for storage and a grid-connected inverter to process power to the utility grid.

A small-scale wind power system is connected with Electrical utility grid by power electronic system that used for interfacing variable speed small wind generators to the utility grid. A small wind turbine consists of permanent magnet synchronous generator (PMSG), AC/DC converter, DC/DC converter as the maximum power point tracking controller, inverter, and load.

6.7.1 AC/DC Converters (Rectifiers)

AC/DC converters are called rectifier circuits, which can be broadly classified based on the type of AC source, and the type of semiconductor switch that is used as shown in Fig. 6.10.

Rectifiers are usually used in either single-phase or three-phase applications, but multi-phase topologies are possible for high-power applications. Switches that are used in rectifier's designs can be either uncontrolled (diode) or controlled devices (thyristor). Single-phase rectifiers can be classified as either half-wave or full-wave circuits.

A three-phase bridge rectifier is commonly used in wind power applications. This is a full-wave rectifier and gives six-pulse ripples on the output voltage. Each one of six diodes conducts for 120°. The pair of diodes which are connected between that pair of supply lines having the highest amount of instantaneous line-to-line voltage will conduct [143]. The three-phase bridge rectifier is shown in Fig. 6.11.

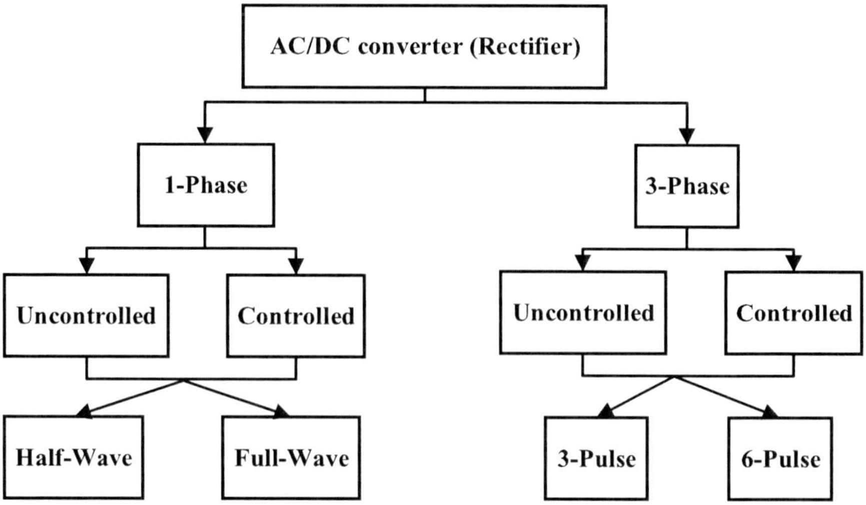

Fig. 6.10 Classifications of rectifier circuits

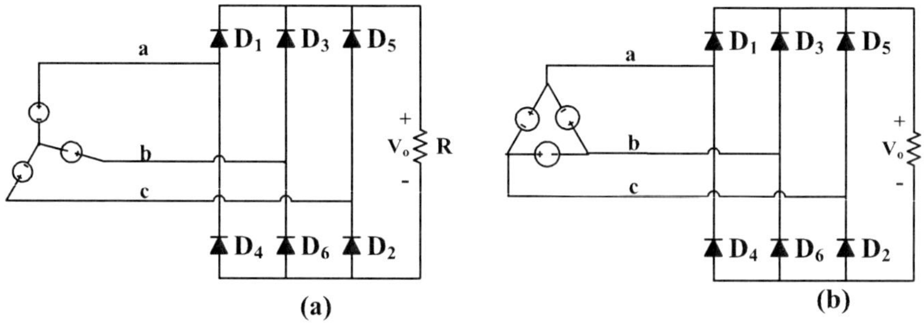

Fig. 6.11 6-pulse rectifier circuits: (**a**) Y-connected source and (**b**) Δ-connected source

If V_m is the peak value of the phase voltage, then the average and RMS output voltage are calculated with the following equation [144]:

$$V_{dc} = V_o = \frac{2}{2\pi/6} \int_0^{\pi/6} \sqrt{3} V_m \cos \omega t \, d(\omega t) = \frac{3\sqrt{3}}{\pi} V_m = 1.654 V_m \tag{6.36}$$

The grid-connected inverters will inject the active power to the grid with minimum total harmonic distortion (THD) [182] of output current and voltage. The grid voltage and inverter output voltage will be synchronized by the zero-crossing circuit. The generator can be a self-excited asynchronous generator (SEIG), or permanent magnet synchronous generator (PMSG). The stator windings are connected to the load or to the grid through a full-scale power converter. Some variable speed WTSs are gearless. In these cases, a direct driven multi-pole generator is used.

A permanent magnet generator has no excitation control, and the output voltage is proportional to the rotor speed. Therefore, in control of a wind turbine, the rotor speed is obtained via the output voltage measurement. The earliest and still most widely used power electronic circuit for this application uses an AC/DC/AC technology in which the variable frequency [183], variable voltage from the generator is first rectified to DC, and then it controlled by DC/DC converter for feeding DC loads and stored energy. Also it converted to AC and fed the grid and AC loads. The continuous variation of wind speed will result in a DC-link voltage varying in an uncontrolled manner. In order to get variable speed operation and stable dc bus voltage, a boost dc-dc converter could be inserted in the DC link. As there is active power flows unidirectional from the PMSG to the DC link through a power converter, only a simple diode rectifier can be applied to the generator side converter in order to obtain a cost-efficient solution [184]. Inverters are studied previously in Chap. 5.

A small wind conversion system may be integrated into loads or power systems with full rated power electronic converters. The wind turbines with a full-scale power converter between the generator and load give the other technical performance. Usually, a back-to-back voltage source converter (VSC) is used to achieve full control of the active and reactive power. But in this case, the control of the whole system would be a difficult task. Since the generator has been decoupled from electric load, it can be operated at a wide range frequency (speed) condition and maximum power extract. Figure 6.12 shows the most used solution with full-scale power converters. This solution has controllable characteristics since the generator is decoupled from the load by a DC link [184].

ANN is used to determine D value depending on inputs of buck-boost converter, input current (I_w) and input voltage (V_w).

Proposed on-grid wind energy system is displayed in Fig. 6.13.

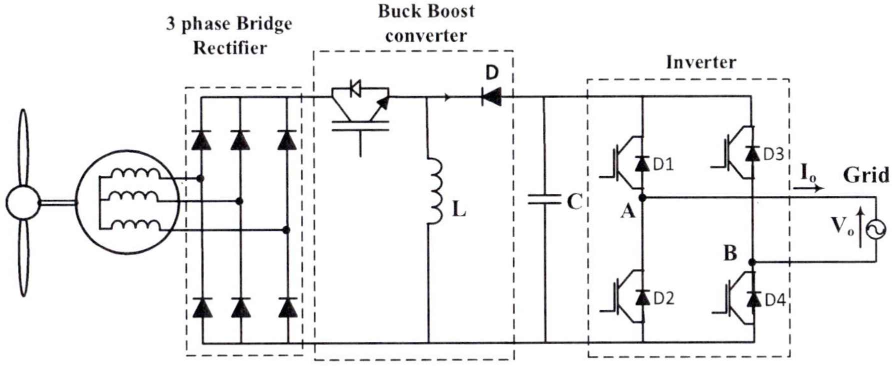

Fig. 6.12 Small-scale wind energy conversion system

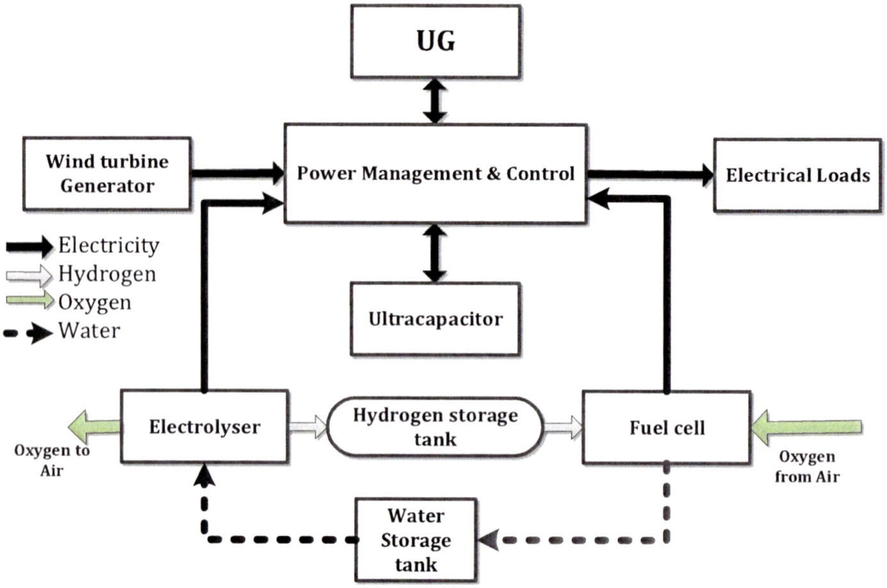

Fig. 6.13 On-grid wind energy system for feeding communication station

6.8 Results and Discussions

Three sites in Egypt are selected for installing a communication station with specific hourly loads that can be shown in Fig. 5.18. The three locations are Shark El-Oweinat, Minia city, and Marsa Alam. Selected WTG with a suitable power rate is used to determine the most suitable for each location. Selected WTG data are shown in Table 6.2.

The wind speeds and weather data from [185] and WTGs data are used to determine Weibull Parameters with Matlab functions. Then the outputs are Weibull parameters C and K for each site and each WTG.

Tables from 6.3, 6.4, and 6.5 display Weibull parameters *C* and *K* for the selected sites under the condition of using the chosen types of WTG. It can be concluded that *K* parameter is a constant value for each location at all hub heights of WTG. *K* parameter equals 3.3101 for Oweinat Site, 2.5283 for Minia site, and 2.5901for Marsa Alam site. *C* parameter values are directly proportional to the hub height of WTG. Pitchwind (H62) WTG has the maximum values of C parameter for all locations ($C = 8.9121$ for Shark El-Oweinat, 4.7147 for Minia, 6.4339 for Marsa Alam) because it has the maximum hub height of the selected WTGs ($H = 62$ m). Gaia and Jonica Impianti WTGs types have the minimum values of *C* parameter (7.4687 for Shark El-Oweinat, 3.9512 for Minia, 5.392 for Marsa Alam) because they have the minimum height of the selected WTG ($H = 18$ m).

Table 6.2 The data of selected WTGs [186]

#	MODS	P_{ewr} (kW)	H (m)	D (m)	u_c (m/s)	u_r (m/s)	u_f (m/s)	η_g (%)
1	Aeolos-V	3	20	2.4	2.5	12	55	96
2	Iskra	5	30	5.4	3	11	60	95
3	Sirocco	6	30	5.6	4	12	None	97
4	Aircon	10	30	7.1	2.5	11	32	96
5	Gaia	11	18	13	3	10	25	97
6	Jonica Impianti	20	18	8	3.5	12.5	37.5	95
7	Eoltec	25	32	10	3	12	None	96
8	Fuhrländer (FL)	30	27	13	2.5	12	25	95
9	Pitchwind (H20)	30	20	14	2	15	None	97
10	Pitchwind (H62)	30	62	14	2	15	None	96
11	FL 100	100	35	21	2.5	13	25	95

Table 6.3 The Weibull parameters, CF and ANWTG for different types of WTGs For Shark El-Oweint

#	MODS	C	K	CF	ANWTG
1	Aeolos-V	7.582	3.3101	0.2122	4
2	Iskra	8.0341	3.3101	0.3237	2
3	Sirocco	8.0341	3.3101	0.2402	2
4	Aircon	8.0341	3.3101	0.3277	1
5	Gaia	7.4687	3.3101	0.3413	1
6	Jonica Impianti	7.4687	3.3101	0.1694	1
7	Eoltec	8.1086	3.3101	0.2588	1
8	Fuhrländer (FL)	7.582	3.3101	0.2122	1
9	Pitchwind (H20)	7.4687	3.3101	0.0983	1
10	Pitchwind (H62)	8.9121	3.3101	0.1768	1
11	FL 100	8.213	3.3101	0.2131	1

Table 6.4 The Weibull parameters, CF and ANWTG for different types of WTGs for Minia Site

#	MODS	P_{ewr} (kW)	C	K	CF	ANWTG
1	Aeolos-V	3	4.0111	2.5283	0.0472	18
2	Iskra	5	4.2503	2.5283	0.0620	7
3	Sirocco	6	4.2503	2.5283	0.0328	13
4	Aircon	10	4.2503	2.5283	0.0712	4
5	Gaia	11	3.9512	2.5283	0.0610	4
6	Jonica Impianti	20	3.9512	2.5283	0.0271	5
7	Eoltec	25	4.2896	2.5283	0.0510	2
8	Fuhrländer (FL)	30	4.1868	2.5283	0.0542	2
9	Pitchwind (H20)	30	4.0111	2.5283	0.0302	2
10	Pitchwind (H62)	30	4.7147	2.5283	0.0481	2
11	FL 100	100	4.3449	2.5283	0.0497	1

Table 6.5 The Weibull parameters, CF and ANWTG for different types of WTGs for Marsa Alam

#	MODS	P_{ewr} (kW)	C	K	CF	ANWTG
1	Aeolos-V	3	5.4737	2.5901	0.1168	8
2	Iskra	5	5.8001	2.5901	0.1636	4
3	Sirocco	6	5.8001	2.5901	0.1100	4
4	Aircon	10	5.8001	2.5901	0.1729	2
5	Gaia	11	5.392	2.5901	0.1682	2
6	Jonica Impianti	20	5.392	2.5901	0.0849	2
7	Eoltec	25	5.8539	2.5901	0.1340	1
8	Fuhrländer (FL)	30	5.7135	2.5901	0.1322	1
9	Pitchwind (H20)	30	5.4737	2.5901	0.0686	2
10	Pitchwind (H62)	30	6.4339	2.5901	0.1069	1
11	FL 100	100	5.9293	2.5901	0.1192	1

A computer program has been designed to calculate the capacity factor, CF, and the average number of each selected WTG type, ANWTG. The input data of this program are:

1. The characteristics of each WTG shown in Table 6.2.
2. The pre-calculated Weibull parameters C and K of each site for all WTGs, which can be shown in tables from 6.3, 6.4, and 6.5.
3. The average load demand.

The outputs of this program are the CF for each WTG, and the average number of each WTG type needed. Table 6.3 display the results for Shark El-Oweinat site. Gaia WTG type has the best performance because it has the largest CF (CF = 0.3413); this means Gaia WTG type is the best suitable WTG from all selected WTG types for Oweinat site. ANWTG of Gaia WTG type required for Shark El-Oweinat site is 1 (ANWTG = 1).

Table 6.4 revels theses results for Minia site. Aircon WTG type has the best performance because it has the largest CF (CF = 0.0712); this means Aircon WTG type is the best suitable WTG from all selected WTG types for Minia site. ANWTG of Aircon WTG type required for Minia site is 4 (ANWTG = 4).

Table 6.5 revels theses results for Marsa Alam site. Aircon WTG type has the best performance because it has the largest CF (CF = 0.1729); this means Aircon WTG type is the best suitable WTG from all selected WTG types for Marsa Alam site. ANWTG Aircon WTG type required for Marsa Alam site is 2 (ANWTG = 2) for Gaia WTG type.

Figures from 6.14, 6.15, and 6.16 display CF of selected WTG for selected sites.

Generally, Oweinat site is the best site for generating electricity from wind because it has the largest CF, (CF = 0.3413). Marsa Alam is the second site for generating electricity from wind because it has CF slightly lower (CF = 0.1729). Minia site is the worst site for generating electricity from wind because it has the smallest CF, (CF = 0.0712), so, Minia site is poor in wind energy.

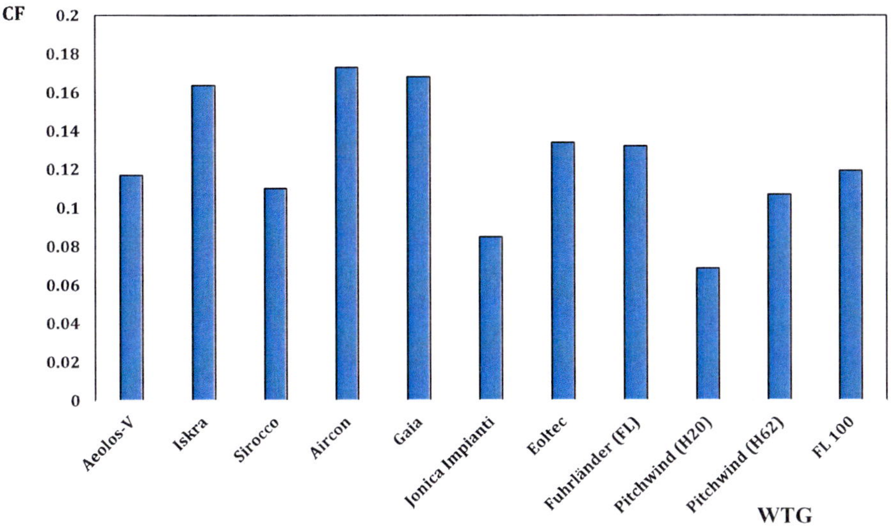

Fig. 6.14 The capacity factor for selected WTG (Shark El-Oweinat)

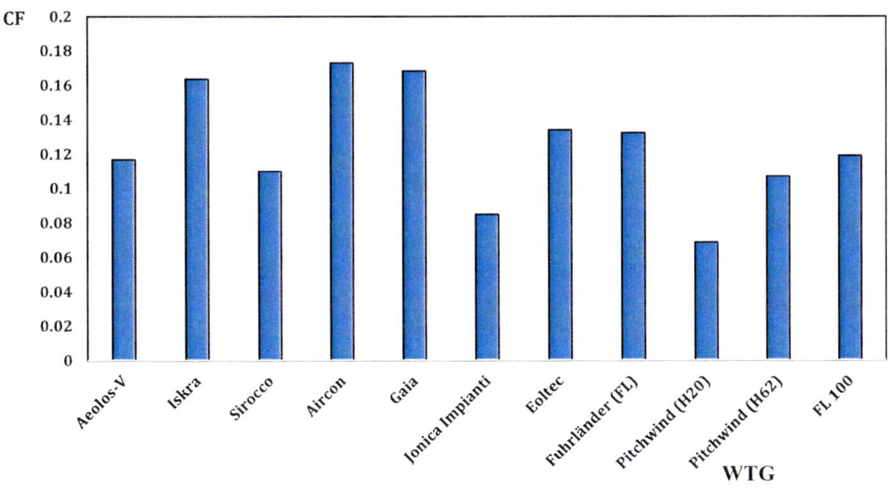

Fig. 6.15 The capacity factors for selected WTG (Minia site)

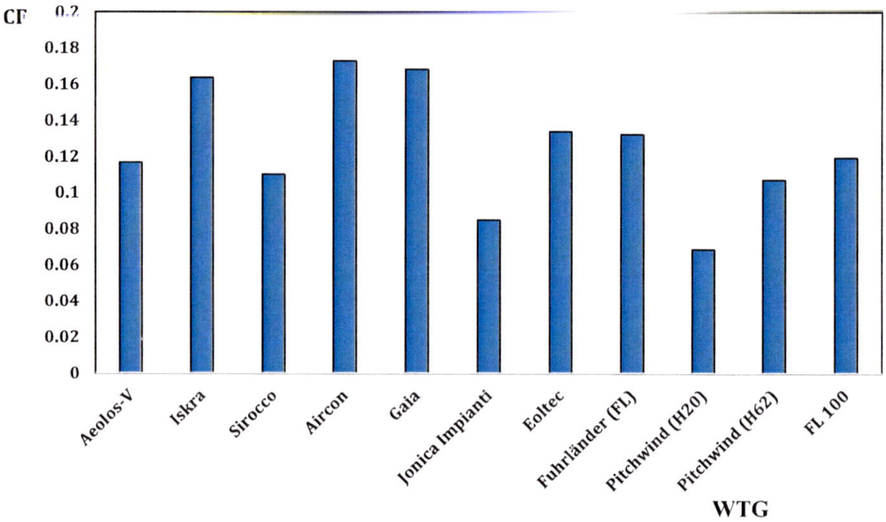

Fig. 6.16 The capacity factors for selected WTG (Marsa Alam)

Table 6.6 ONWTG, yearly generated energy and storage size required

Site	Mods	ONWTG	YE, kWh	Storage size required, kWh
Shark El-Oweinat	Gaia	1	32325.8293	488.7594
Minia	Aircon	4	24808.5077	2721.9054
Marsa Alam	Aircon	2	30535.5135	760.8269

A new computer program is designed for carrying the energy balance study discussed in Sect. 6.2. The inputs of this program are:

1. The hourly wind speed of each site. This data is taken from Ref [53].
2. The hourly electrical load from Fig. (5.18).
3. The characteristics of the selected WTG that match every site. These characteristics are P_{ew}, u_c, u_r, u_f, and H. Table 6.2 displays these characteristics.
4. The Weibull parameters (C and K) and the average number of WTG, ANWTG, for the selected WTG that match every site that can be shown in from Tables 6.3, 6.4, and 6.5.

The outputs of this program are:

1. The hourly generated power.
2. The optimum number of WTG, ONWTG, and the yearly output energy that are shown in Table 6.6.
3. Surplus and deficit energy every month that can be displayed in tables from 6.7, 6.8, and 6.9 and in figures from 6.17, 6.18, and 6.19.
4. Maximum deficit and maximum surplus over all the year.

Table 6.7 Surplus/deficit energy (kWh) during all year months in Sharl El-Oweinat using Gaia WTG

Month	Jan.	Feb.	Mar.	Apr.	May	Jun.	Jul.	Aug.	Sep.	Oct.	Nov.	Dec.
dE	−488.8	185.1	804.9	170	−40.6	1501.8	1305.9	1140.1	2491.9	1647	207.9	1369.5

Table 6.8 Monthly surplus/deficit energy (kWh) for Minia (Aircon Mod, $N = 4$)

Month	Jan.	Feb.	Mar.	Apr.	May	Jun.	Jul.	Aug.	Sep.	Oct.	Nov.	Dec.
dE	−760.7	923.7	521.1	906.2	1928.8	80.14	704.1	334.6	−55	155.91	−810.7	−1150.5

Table 6.9 Monthly surplus/deficit energy (kWh) for Marsa Alam site (Aircon Mod, $N = 2$)

Month	Jan.	Feb.	Mar.	Apr.	May	Jun.	Jul.	Aug.	Sep.	Oct.	Nov.	Dec.
dE	−291.1	−415.7	1790.9	3169.6	765.9	1151.91	235.1	777.7	878.35	411.75	84.2	−54

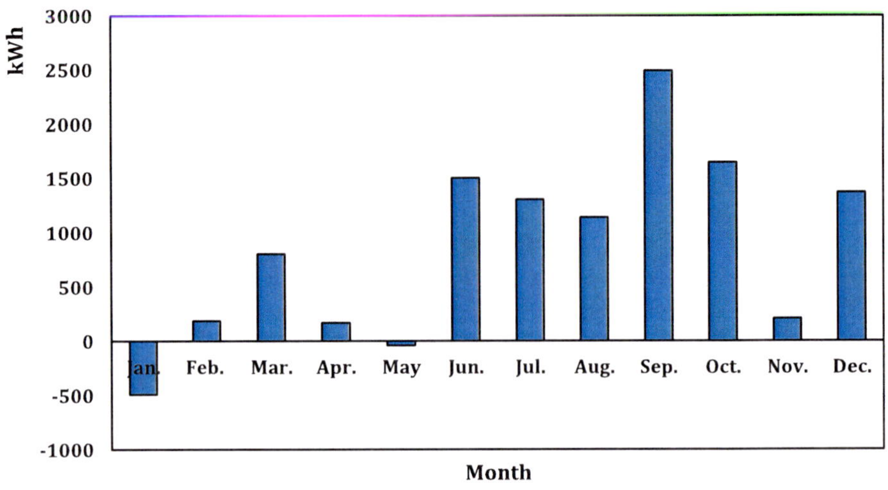

Fig. 6.17 Surplus/deficit energy during all year months in Shark El-Oweinat

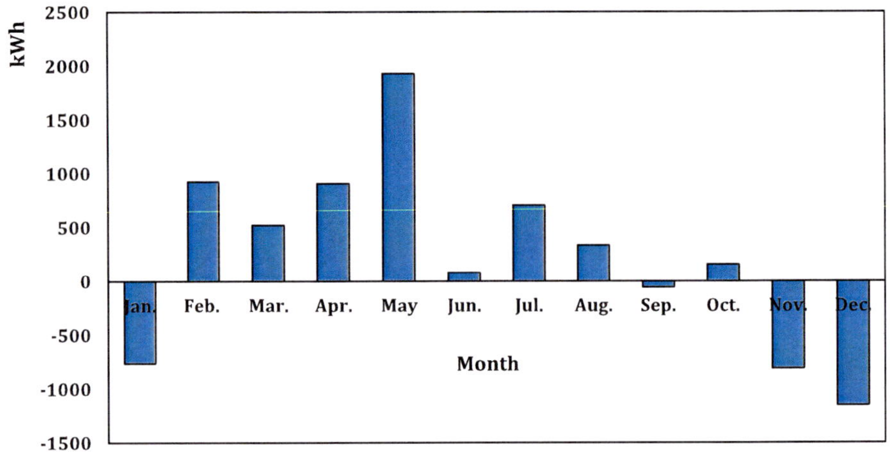

Fig. 6.18 Monthly surplus/deficit energy (kWh) for Minia (Aircon Mod, $N = 4$)

5. Energy storage size required for every site.
   Max. Deficit = sum deficit in Nov, Dec, and Jan = 2721.91 kWh.
   Hyd tank size must be sufficient to generate energy equal to 2721.91 kWh.

   Max. Deficit = sum deficit in Nov, Dec, and Jan = 760.8269 kWh.
   Hyd tank size must be sufficient to generate energy equal to 760.8269 kWh.

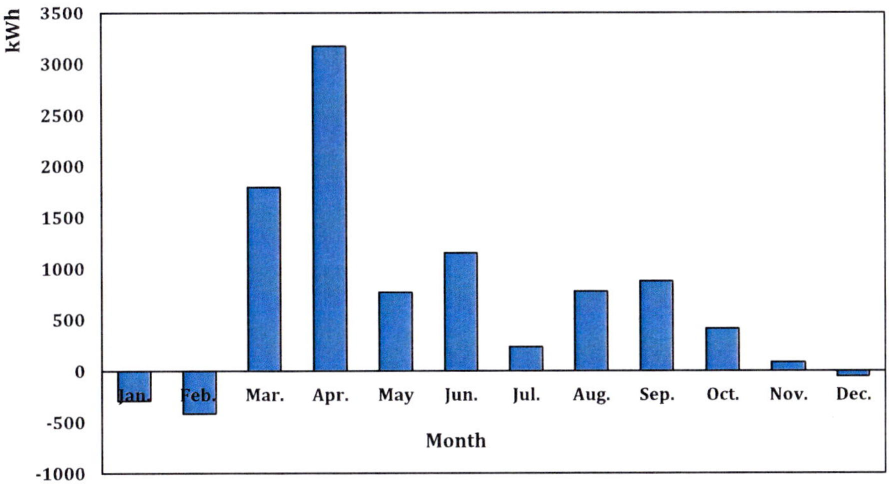

Fig. 6.19 Monthly surplus/deficit energy (kWh) for Marsa Alam site (Aircon Mod, $N = 2$)

Table 6.10 Hydrogen Prod., Usage, and surplus/deficit every month in Shark El-Oweinat site

Month	Hyd. Prod. (N m^3)	Hyd. Usage (N m^3)	Surp/Def. (N m^3)
Jan.	133.3346	406.646	−273.311
Feb.	361.182	252.7084	108.4736
Mar.	690.85694	229.7523597	461.1046
Apr.	302.352	203.072	99.2804
May	277.71	297.044	−19.334
Jun.	961.6719	105.246	856.4259
Jul.	860.1656	115.1597	745.0059
Aug.	864.5196	212.7472	651.7723
Sep.	1660.357	238.5569	1421.8
Oct.	1335.879	393.2558	942.6234
Nov.	575.172	451.253	123.919
Dec.	1143.091	358.9029	784.1877

Regenerative fuel cells system is used as a backup supply and mean of stored energy. Chapter 3 displays the methodology of the RFC system design. By applying this methodology and specs of RFC in Chap. 5, the following results are getting:

For Shark El-Oweinat site Hydrogen Prod., Usage, and surplus/deficit in, N m^3, every month can be shown in Table 6.10 and Figs. 6.20 and 6.21.

For Marsa Alam site Hydrogen Prod., Usage, and surplus/deficit in, N m^3, every month can be shown in Table 6.11 and Figs. 6.22 and 6.23.

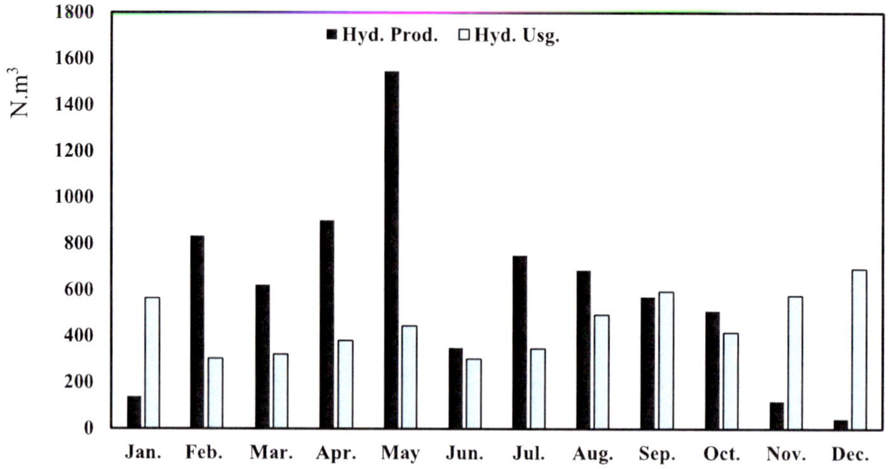

Fig. 6.20 Hydrogen Prod., Usage in Shark El-Oweinat site

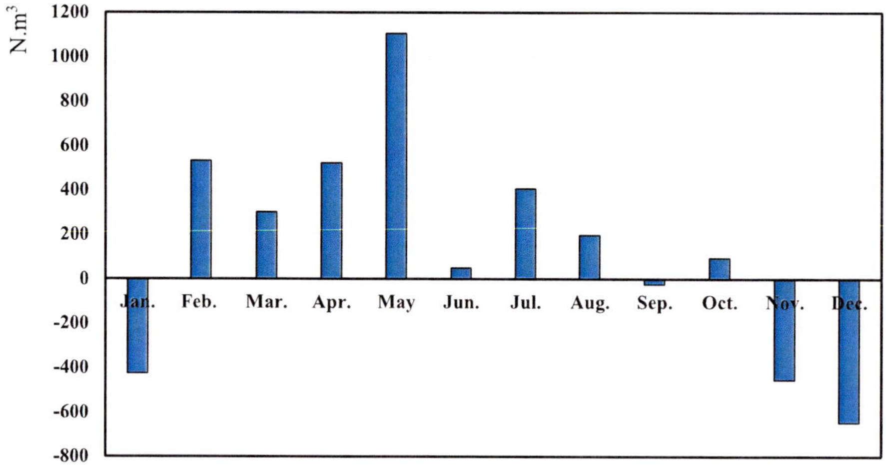

Fig. 6.21 Hydrogen surplus/deficit every month in Shark El-Oweinat

For Minia site Hydrogen Prod., Usage, and surplus/deficit in, N m^3, every month can be shown in Table 6.12 and Figs. 6.24 and 6.25.

In our locations the generated energy from wind turbines is converted from AC to DC using rectifier then DC-DC converter is used to have the required voltage and current to feed DC loads and store surplus energy. For UG connections DC to AC inverter is connected as shown in Figs. 6.26 and 6.27.

Buck-boost DC-DC converter is used. Table 6.13 displays the details of the design of WES in our selected sites. The design details depend on wind speeds, by

Table 6.11 Hydrogen Prod., Usage, and surplus/deficit, N m^3, every month in Marsa Alam

Month	Hyd. Prod. (N m^3)	Hyd. Usage (N m^3)	Surp/Def. (N m^3)
Jan.	319.3988	479.2626	−159.864
Feb.	204.8626	436.2297	−231.367
Mar.	1280.568596	257.6808	1022.888
Apr.	2021.47	214.058	1807.41
May	673.434	234.444	438.991
Jun.	914.2577	255.2298	659.028
Jul.	576.871	437.6532	139.2178
Aug.	838.9947	391.3811	447.6136
Sep.	773.8905	270.4387	503.4519
Oct.	524.0012	286.04	237.9612
Nov.	388.165	336.103	52.0623
Dec.	309.3374	335.9717	−26.6342

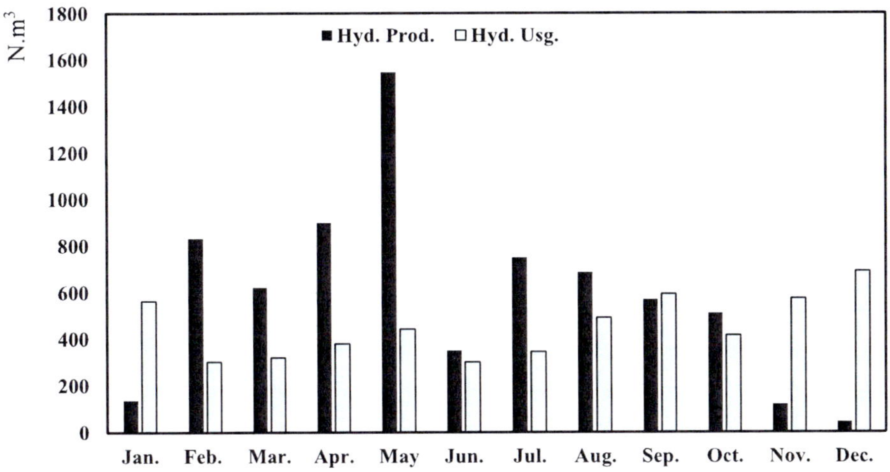

Fig. 6.22 Hydrogen Prod., Usage in Marsa Alam

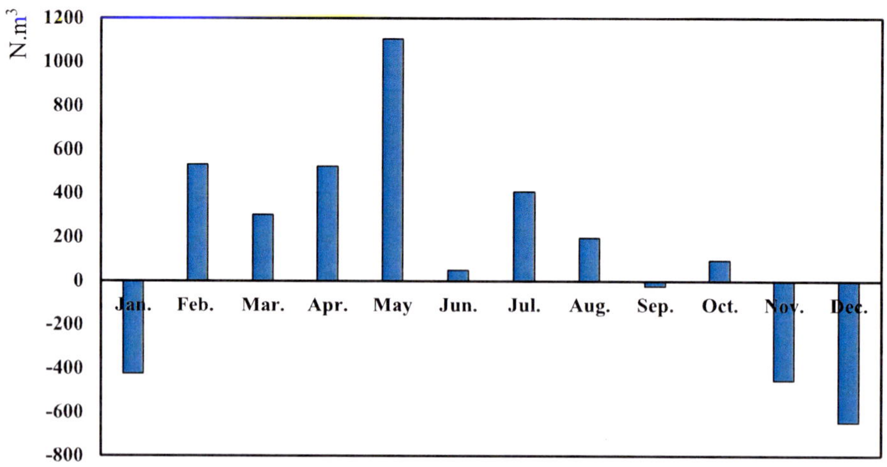

Fig. 6.23 Hydrogen surplus/defect every month in Marsa Alam

Table 6.12 Hydrogen Prod., Usage, and surplus/deficit every month in Minia

Month	Hyd. Prod. (N m^3)	Hyd. Usage (N m^3)	Surp/Def. (N m^3)
Jan.	137.0879	563.3157	−426.228
Feb.	832.1244	302.4763	529.6481
Mar.	621.1017153	320.4604	300.6413
Apr.	899.421	378.77	520.651
May	1545.62	441.964	1103.65
Jun.	349.3119	299.998	49.31388
Jul.	750.0057	344.8729	405.1328
Aug.	686.8491	490.3153	196.5338
Sep.	569.3524	593.3936	−24.04119
Oct.	508.7225	414.86	93.86252
Nov.	119	573.602	−454.602
Dec.	44.24432	690.8465	−646.602

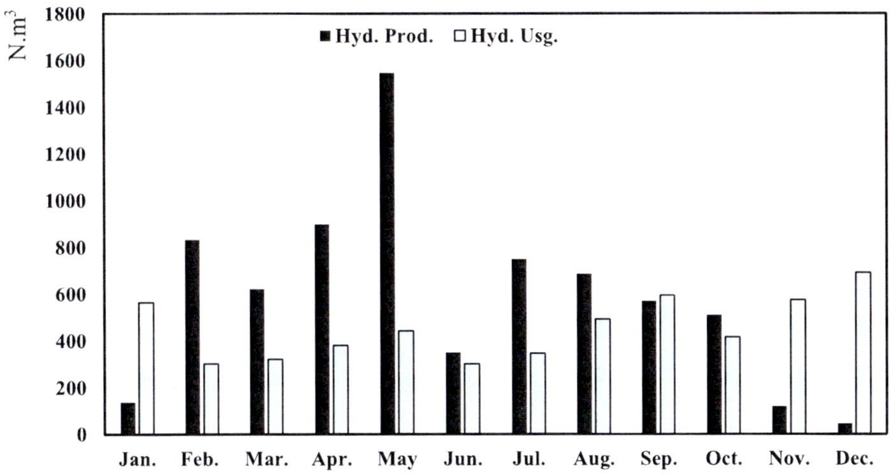

Fig. 6.24 Hydrogen Prod., Usage in Minia

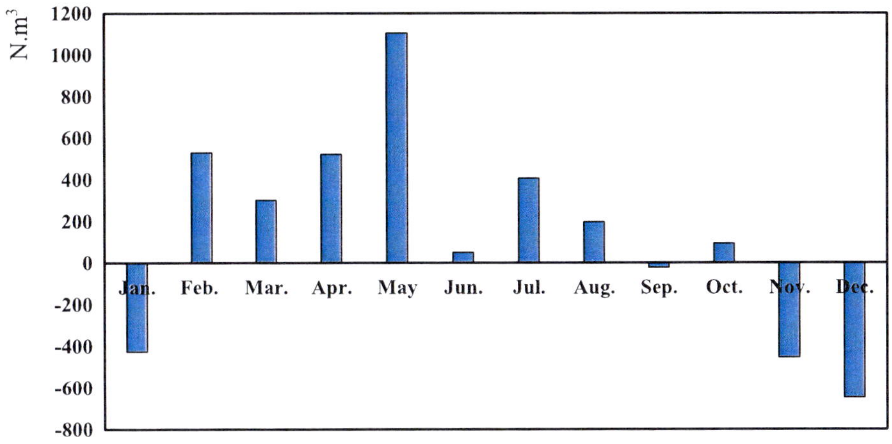

Fig. 6.25 Hydrogen surplus/defect every month in Minia

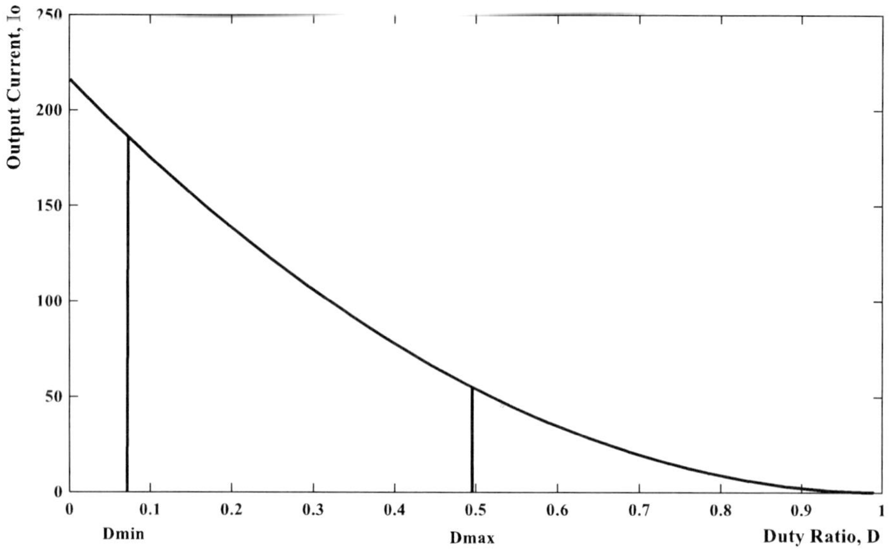

Fig. 6.26 Duty ratio vs. output current with min. and max. duty ratio for buck-boost converter for Gaia WTG in Shark El-Oweinat

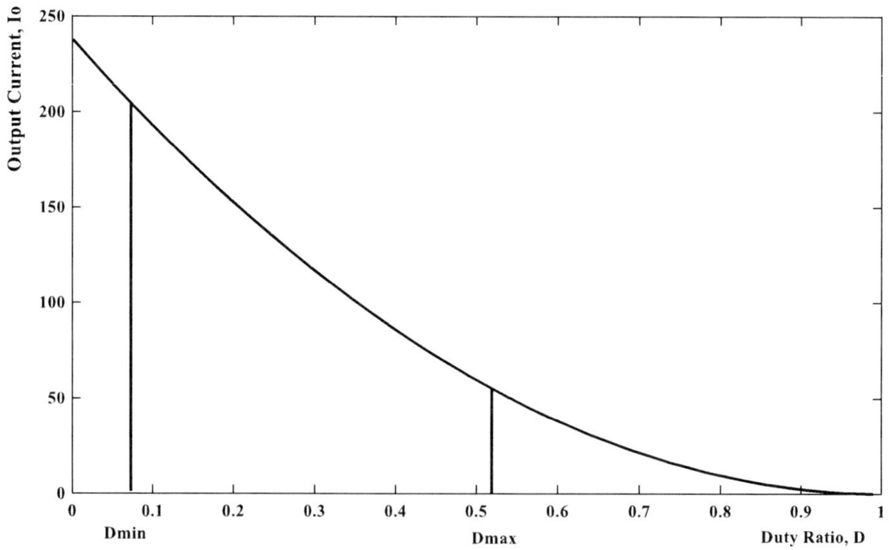

Fig. 6.27 Duty ratio vs. output current with min. and max. duty ratio for buck-boost converter for Aircon WTG in Marsa Alam and Minia sites

Table 6.13 Design data of WES for our selected location

Location	Shark El-Oweinat	Marsa Alam	Minia
WTG	Gaia	Aircon	Aircon
ONWTG	1	2	4
P_{ewr}, (kWh)	11	10	10
V_{ewr}, (V)	400	400	400
V_{rms}, (V)	661.6	661.6	661.6
V_o, (V)	54	54	54
D_{min}	0.07546	0.07546	0.07546
D_{min}	0.51717	0.49355	0.49355
L, (μH)	7.5533	8.31	8.31
C, (mF)	1.8976	1.7252	1.7252

Table 6.14 The weights and bases of the input layer (2 + 10 + 1) ANN

Weights		Base
−1.8461	−11.4089	−10.0559
4.4656	−2.6331	−6.7984
−12.4386	12.259	25.7247
6.9045	−0.24789	−6.4674
0.81191	−3.5585	−1.2906
−1.0356	3.8122	−0.97629
−0.69982	4.1309	2.7896
−3.8642	−0.77748	−2.6372
−3.4508	1.7566	−5.5779
−3.7477	−2.3374	−4.4492

Table 6.15 The weights and bases of the output layer (2 + 10 + 1) ANN for determining the duty ratio

Weights output									
12.1684	−2.8009	9.2646	0.91645	0.14243	0.077432	−0.19013	−0.80715	−1.7515	1.2214
Bases output									
−0.60655									

applying hourly electrical loads in fig. (5.18), and data of the selected WTG from item 6.8. The designed data will be as follows (Tables 6.14 and 6.15):

6.8.1 ANN Results for Determining Duty Cycle

ANN architecture (2 + 10 + 1) is designed for determining duty cycle in Shark El-Oweinat location because this location is the optimum for using wind energy

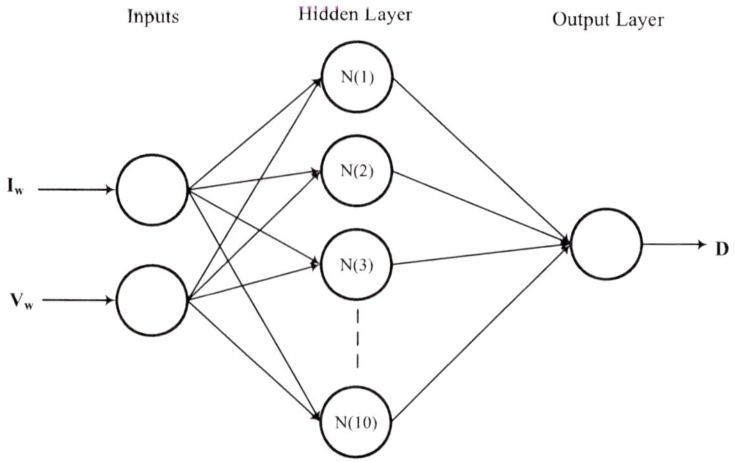

Fig. 6.28 ANN architecture (2 + 10 + 1)

system related to other location in this study. As shown in Fig. 6.28, the inputs are I_w and V_w, the output is value of duty cycle for extracting maximum power. ANN Matlab tool is used for doing this task. Figure 6.29 displays neural networks training tool. Validation performance is shown in Fig. 6.30. The training test results are displayed in Figs. 6.31 and 6.32.

Figure 6.33 displays the outputs from analytic method and ANN outputs. It can be noted that there are very small errors between outputs and ANN outputs.

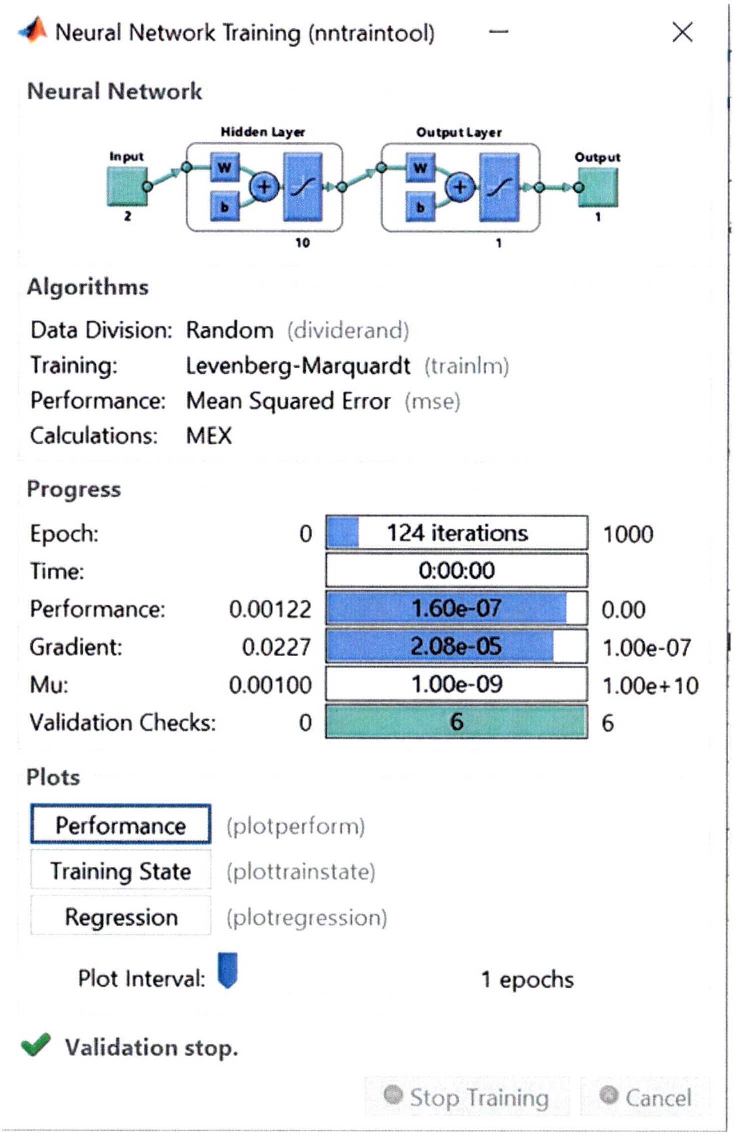

Fig. 6.29 ANN data training

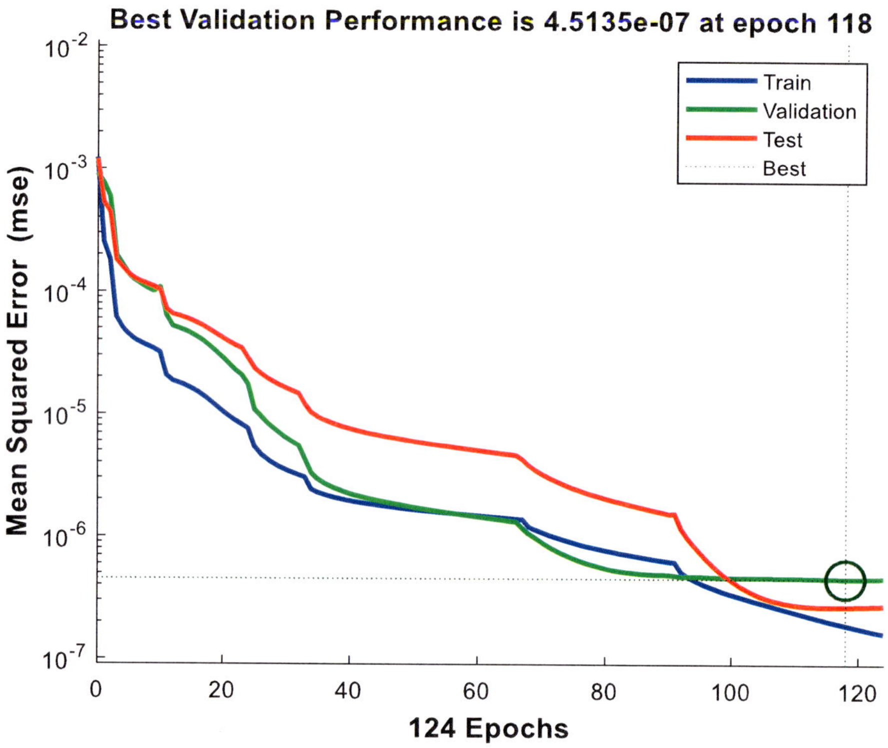

Fig. 6.30 Validation performance for ANN (3 + 10 + 1)

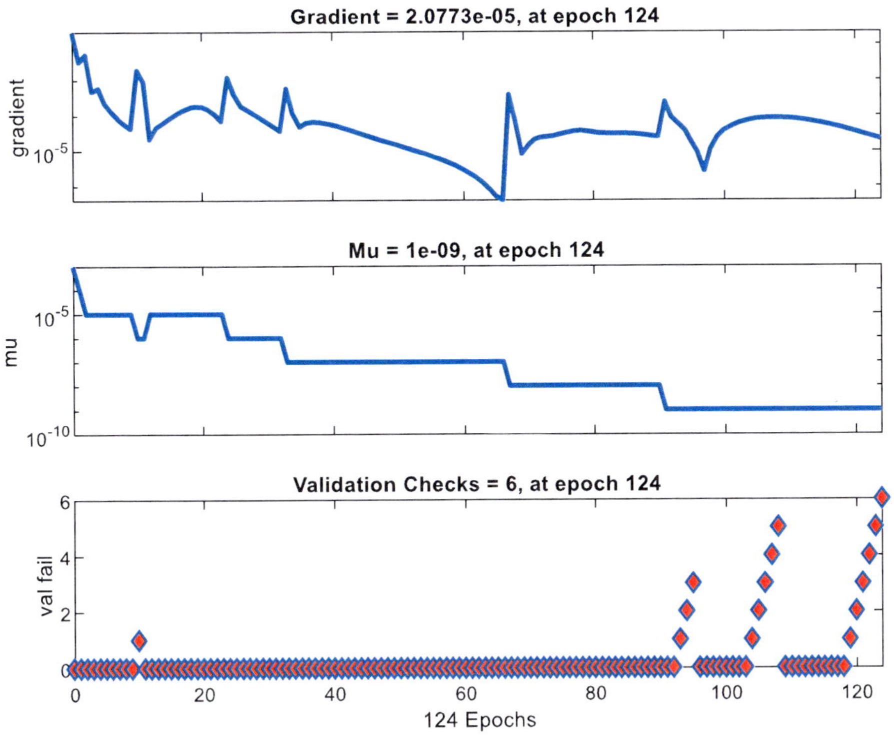

Fig. 6.31 The training test results of ANN (2 + 10 + 1)

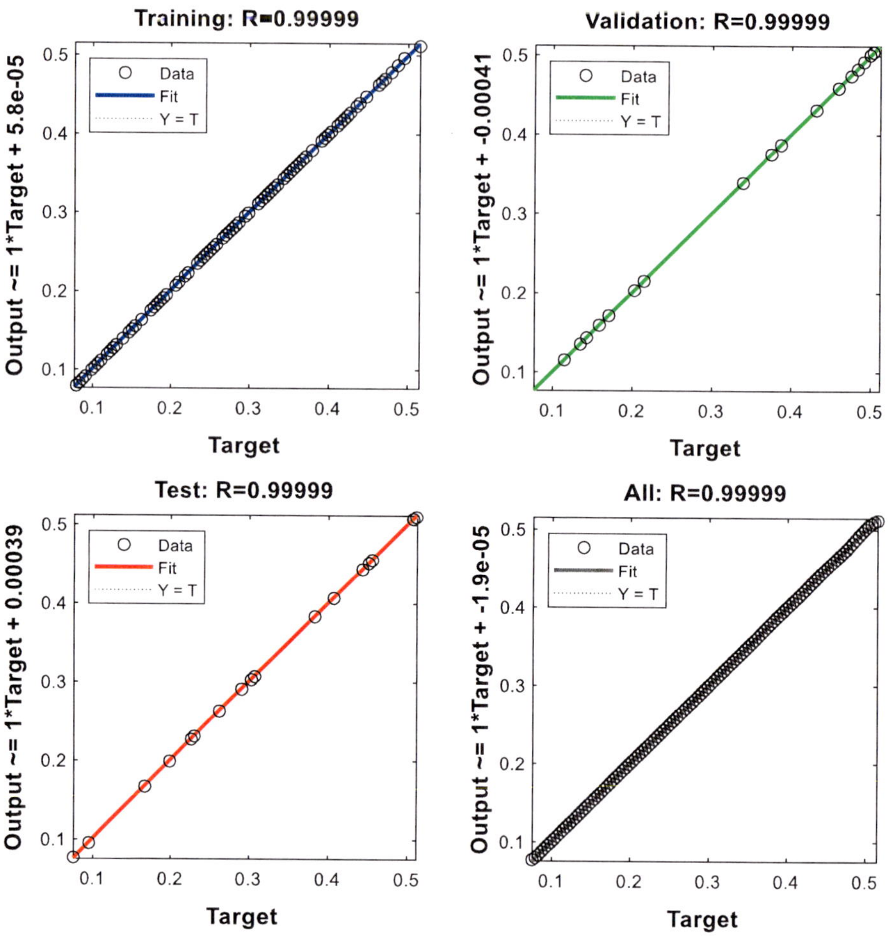

Fig. 6.32 The regression of ANN (2 + 10 + 1)

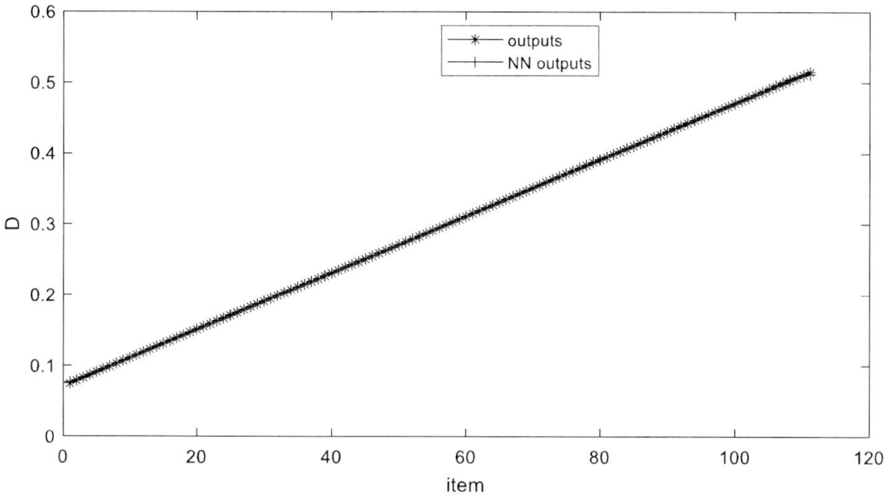

Fig. 6.33 Comparison between outputs and NN outputs

6.9 Summary

This chapter explores the wind energy power system. This system contains wind turbines as the main source and RFC as a backup source and means of stored energy in the form of hydrogen. Equations have been provided for the calculation of generated wind energy. Eleven types of wind turbines with appropriate capacity rates for the loads were selected to choose the best of them according to wind speeds for each of the sites selected for the study also according to the value of the power coefficient of each site, where it is choosing the highest power coefficient because this means that it is the most appropriate for this site. A capacity balance program has also been developed to determine the optimal number of wind turbines required to feed loads, taking into account the efficiency of RFC. The value of hydrogen produced by energy conversion was determined during the surplus times. The value of hydrogen consumption during defect times which need to be equal hydrogen production. The connection of the wind energy system to the UG has also been studied. ANN technique is used to obtain maximum power.

Chapter 7
Optimum Sizing and Design of Renewable Energy System

7.1 Introduction

In spite of the great advantages of renewable energies opposed to conventional energies, they share common weaknesses, like the great suffering to the weather, the poor stability, and the heavy randomness, which leads to the poor reliability of generation and low efficiency. Therefore, the hybrid renewable energy system can take care of the main problems and limitations in terms of the efficiency, reliability, and the economics, which can be considered an important solution to fulfill load demands, support and improve the system [122]. Most of the researchers have optimized the size, design, control and economic analysis for hybrid renewable systems, which implemented different renewable sources like a wind turbine or/and solar module or/and other renewable energy sources with or without energy storage systems [187]. To overcome the problems of hybrid systems and achieve the objectives of its establishment, it has been applied to various methods, such as optimization techniques, software packages, and analytical techniques.

The hybrid power generation concept is a system aimed at the production and utilization of electrical energy coming from more than one source within an integrated arrangement. The hybrid system studied in this book is one combining solar PV and wind turbines with power conditioning units such as inverters. Hybrid wind turbine and solar PV modules offer more excellent reliability than any one of them alone because local energy supply cannot depend entirely on any one of these sources. Other advantages of the hybrid system are the stability and reliability of the system and the lower maintenance requirement, thus reducing downtime during repairs and routine maintenance. In addition to this, as well as being indigenous and free, renewable energy resources also contribute to the reduction of pollution emissions.

In this book, the proposed hybrid power generation system makes use of solar PV and wind turbine to produce electricity and supply the load by connecting to the grid.

A. A. Elbaset, S. Ata, *Hybrid Renewable Energy Systems for Remote Telecommunication Stations*, https://doi.org/10.1007/978-3-030-66344-5_7

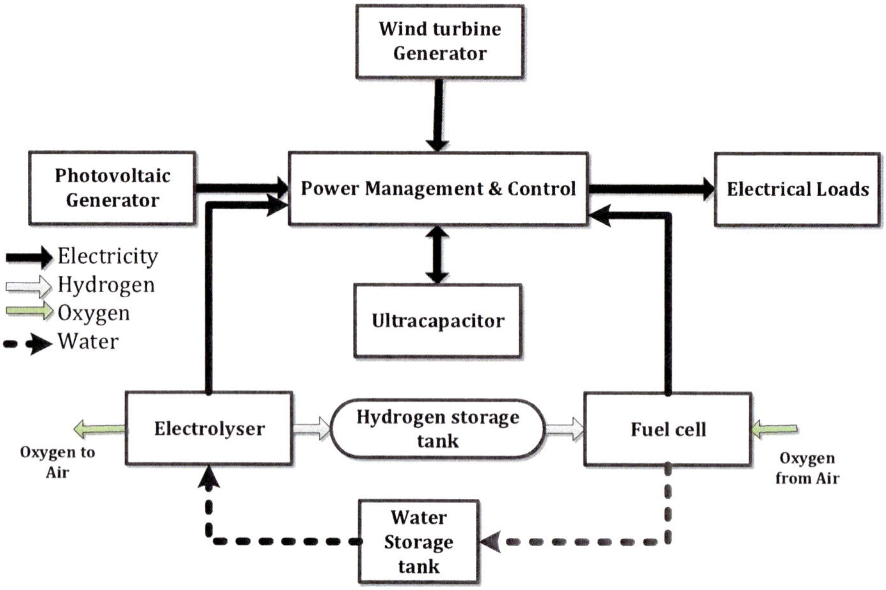

Fig. 7.1 Hybrid PV/WT/RFC/UC power system

A schematic of a typical grid-connected wind-solar hybrid system can be shown in Fig. 7.1.

Many researchers have used different optimization techniques for the hybrid renewable system to reduce the total cost of energy production and improve the efficiency of the hybrid system. As an example, several of these researches harnessed the Evolutionary Algorithm (EA) to consider multi-objective in the optimization issue of a hybrid system [188] (Fig. 7.2).

Authors in [189] designed optimum sizing of hybrid wind turbines/solar panels with the storage system by employing fuzzy adaptive with GA which resolves the optimum number of these components; also, GA is chosen to achieve the optimal power arrangement of a stand-alone hybrid renewable system. In [190], the authors handle the optimal capacity combination of an autonomous hybrid power system consisting of solar panels and wind turbines. The way out of these nonlinear problems is a typical intelligent optimization algorithm. Consequently, the PSO algorithm has been applied and improved to solve this objective; the result achieved proves its effectiveness and performance. In [191] the PSO optimization was presented to fulfill multiple goals in real-time energy management for a hybrid independent system consists of the wind micro-turbine energy system. From all papers mentioned above, it can be argued that the use of hybrid resources makes the optimization an intricate problem, and it becomes difficult to obtain the global optimum solution. To overcome this problem, various artificial intelligence algorithms can apply and develop. Genetic algorithms and PSO are among the most famous intelligent algorithms, they have a good ability in dealing with nonlinear problems and avoid the local minimum

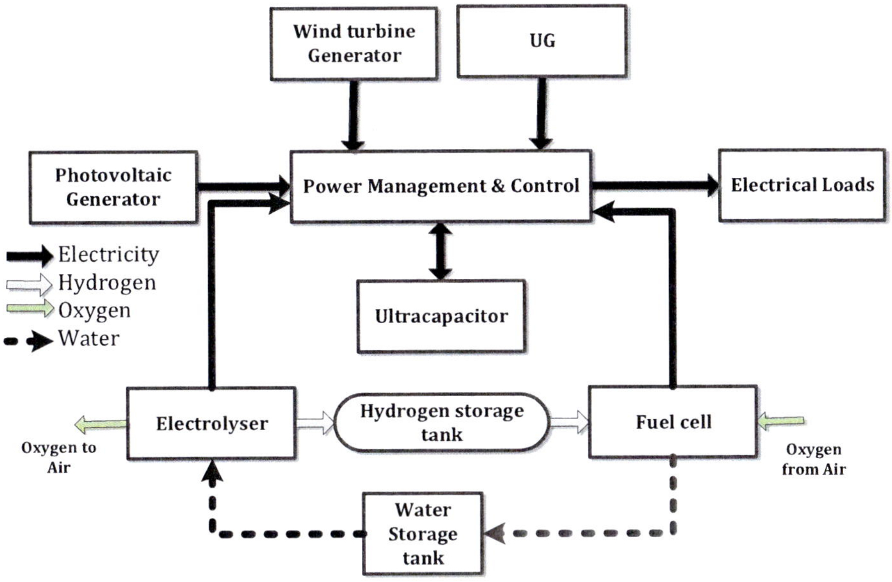

Fig. 7.2 Hybrid PV/WT/RFC/UC/UG power system

solution [192], but the genetic algorithm has main problems represented in it's complicated and it takes a lot of time to achieve the optimal solution. To surmount these obstacles, in this article, the Particle Swarm Optimization technique (PSO) was proposed and developed for optimal performance; the ability of this algorithm was developed to reach the best results by making the particles of the algorithm not only follow the traditional formula of the algorithm, in which the particles search and follow the nearest and best particle, but it was developed to review all particles and look at the best particle in the group and follow it; the optimal solution was achieved by double speed, where it reached the optimal solution at a time rate better than 80% of conventional technology time and less than 20 repetitions only. In addition, this algorithm has been harnessed to achieve multiple objective functions. The stand-alone hybrid renewable energy system depends on solar panels, tidal turbines, wind turbines and with battery bank in order to fulfill an electric load located in the far west of France in Brittany, the results showed the good ability of the PSO developer to solve hybrid system optimization problems such as optimal size, optimal the total cost of energy (COE), and optimal energy management with superior speed and accuracy taking into consideration the several constraints as the State Of Charge (SOC) of the energy storage system, high reliability, and planning expansion for the future development. A variety of different scenarios have been studied to meet the electric load demand. PSO algorithm program was developed using MATLAB software and the total net present cost (TNPSC) is proposed as an objective function.

7.2 Hybrid System Description

The hybrid renewable energy system model it was built in this study depending on solar panels, wind turbines, regenerative fuel cells, ultracapacitor, converters, controller, and other devices were proposed to fulfill a load demand in Fig. 5.18. The design of hybrid WTG/PV/FC/UC power system depends on dividing the load into two parts between PV and WTG. The penetration level parameters determine the partition of load. But PSO technique is used to get the optimization and predict the hybrid system performance, several components must be modeled first as in following.

7.2.1 *The Load Demand Model*

In this study, it is required for the power communication station. The hourly load demand for the communication station is illustrated in Fig. 5.18.

7.2.2 *Solar Panel*

The hourly generated power $P_v(t)$ can be calculated from Sect. 5.2. Then, $P_v(t)$ will be compared with the hourly load demand $P_L(t)$*PELV as follows [194]:

$$\mathrm{BALW} = \sum_{t=1}^{t=8760} [P_v(t) - \mathrm{PELV} \times P_L(t)] \tag{7.1}$$

where PELV is the penetration level of P_V (the ratio of the load demand that fed by P_V).

Depending upon the result of Eq. (7.1):

1. If BALV > 0, then solar cells area, SCA, must be decreased by an incremental area and return calculations.
2. If BALV < 0 then solar cells area, SCA, must be increased by an incremental area and return calculations.
3. If BALV $= 0$ then the computed SCA represent the optimum solar cells area, OSCA, for certain PELV.

7.2.3 *Wind Turbine*

The hourly generated power for WTG $P_{ew}(t)$ can be determined from Sect. 6.2. Then $P_{ew}(t)$ will be compared with the hourly load demand $P_L(t)$*PELW as follows [122]:

$$\text{BALW} = \sum_{t=1}^{t=8760} \left[P_{\text{ew}}(t) - \text{PELW} \times P_{\text{L}}(t)\right] \tag{7.2}$$

where PELW is a penetration level of WTG (the ratio of the total load demand that fed by WTG).

There are three scenarios for BALW; these scenarios can be concluded as following [194]:

1. If BALW > 0, then the number of WTG is larger than the optimum; WTG number must be decreased by one unit and repeat estimation of BAL_W.
2. If BALW < 0, then the number of WTG is smaller than the optimum; WTG number must be increased by one unit and repeat estimation of BAL_W.
3. If BALW $= 0$, then the number of WTG represents the optimum number of WTG units, ONWTG, for certain PELW.

7.2.4 Regenerative Fuel Cell

Extensive study about RFC can be found in Chap. 3 and results of RFC with PV can be found in Chap. 5 and also results of RFC with wind turbine generator can be found in Chap. 6.

7.2.5 Ultracapacitor

Extensive study about basics of UC, its characteristics, different types of sizing, and results can be found in Chap. 4.

7.2.6 Converters and Controllers

Long discussion about converters and controllers and application results of PVPS can be found in Chap. 5; application results of converters in WES can be found in Chap. 6

7.2.7 Energy Management and Operation Strategy

In this study, the energy management process in the hybrid system can be represented in three main cases [122]: The first case if the total energy production by the hybrid system exceeds the electrical load, the surplus energy will be converting to hydrogen by EZ. The second case if the electrical load is higher than hybrid energy production, the FC will feed the load to satisfy the deficit in the system. The third case if the electrical load is equal with the available power production, there is no charge or discharge.

7.3 System Cost Analysis

The aim of sizing the RE system is to obtain the actual size of each system component that will satisfy load demand [195, 196] at minimum cost and subject to some constraints. For the purpose of evaluating the cost objective function, the following costs are considered.

- Capital cost of hybrid system components such as the inverter, H2 tank (HT), WT, FC, EL, and PV module, etc.
- Cost of maintaining the components.
- Cost of replacing components.
- Cost of fuel throughout the entire lifespan of the system (in our system no fossil fuel used).

Different methods were used in evaluating the system cost. It can be based on net present cost (NPC), the annualized cost of the system (ACS), Levelized cost of energy (LEC), and Lifecycle cost (LCC).

7.3.1 Problem Statement

The aim of this section is to introduce an algorithm to solve the problem of sizing of the hybrid renewable energy power system to supply the load demand by considering the minimum cost [197]. In the proposed algorithm, cost estimation is created based on the concept of levelized energy cost (*LEC*) [198]. Cost estimation of the hybrid renewable energy power system is detailed in the following subsections:

7.3.2 Cost Estimation

LEC is a standout among the most well-known and utilized indicator of economic analysis of HRES, and it can be calculated from the following equation [198]:

$$\mathrm{LEC} = \frac{\mathrm{TPC} \times \mathrm{CRF}}{\mathrm{LAE}} \tag{7.3}$$

where TPC is the total present cost of the entire system, LAE is the annual load demand, and CRF is the capital recovery factor. CRF is expressed as shown in the following equation [198]:

$$\mathrm{CRF} = \frac{r \times (1+r)^{\mathrm{LT}}}{(1+r)^{T} - 1} \tag{7.4}$$

where r is the net interest rate and LT is the system lifetime in years.

TPC of entire system is the sum of components present cost, TPC_x; these components form the entire system, and TPC_x of each component can be calculated from the following equation:

$$\mathrm{TPC}_x = \mathrm{IC}_x + \mathrm{OMC}_x + \mathrm{RC}_x \tag{7.5}$$

where IC_x: the initial capital cost of each component, \$. OMC: operation and maintenance cost, \$. RC: replacement cost, \$.

Let's take more details about equipment cost:

Photovoltaic panel cost:

$$\mathrm{TPC}_{\mathrm{pv}} = 1.4 * \mathrm{IC}_{\mathrm{pv}} + \mathrm{OMC}_{\mathrm{pv}} + \mathrm{RC}_{\mathrm{pv}} \tag{7.6}$$

$$\mathrm{TPC}_{\mathrm{pv}} = 1.4 * \mathrm{PVP} * \mathrm{NOP} * \mathrm{CPV} + \mathrm{OMC}_{\mathrm{pv}} + \mathrm{RC}_{\mathrm{pv}} \tag{7.7}$$

where PVP: peak power of each PV panel, kW. NOP: number of PV panels. CPV: capital cost of PV panels, \$/kW.

The capital cost of the PV system is multiplied by 1.4 as in previous equation to include establishment cost (structure cost).

Wind turbine cost:

$$\mathrm{TPC}_{\mathrm{WT}} = 1.23 * \mathrm{IC}_{\mathrm{WT}} + \mathrm{OMC}_{\mathrm{WT}} + \mathrm{RC}_{\mathrm{WT}} \tag{7.8}$$

$$\mathrm{TPC}_{\mathrm{WT}} = 1.23 * \mathrm{WTP} * \mathrm{NWT} * \mathrm{WTC} + \mathrm{OMC}_{\mathrm{WT}} + \mathrm{RC}_{\mathrm{WT}} \tag{7.9}$$

where WTP: peak power of wind turbine, kW. NWT: number of wind turbines. WTC: capital cost of a wind turbine, \$/kW.

The capital cost of the wind turbine is multiplied by 1.23 as in previous equation to include establishment cost (structure cost).

Discrete Regenerative fuel cells cost:

$$\mathrm{TPC}_{\mathrm{DRFC}} = \mathrm{TPC}_{\mathrm{FC}} + \mathrm{TPC}_{\mathrm{EZ}} + \mathrm{TPC}_{\mathrm{H}} \tag{7.10}$$

where $\mathrm{TPC}_{\mathrm{DRFC}}$: total present cost of discrete regenerative fuel cells. $\mathrm{TPC}_{\mathrm{FC}}$: total present cost of the fuel cell part. $\mathrm{TPC}_{\mathrm{EZ}}$: total present cost of electrolyzer part. $\mathrm{TPC}_{\mathrm{H}}$: total present cost of hydrogen and water equipment (tanks and reformers).

Fuel cell part cost can be calculated as follows:

$$TPC_{FC} = IC_{FC} + OMC_{FC} + RC_{FC} \tag{7.11}$$

$$TPC_{FC} = FCR * FCC + OMC_{FC} + RC_{FC} \tag{7.12}$$

where FCR: fuel cell rate, kW. FCC: fuel cell capital cost, \$/kW. OMC_{FC}: operation and maintenance cost of fuel cell, \$/kW. RC_{FC}: Fuel cell replacement cost, \$/kW.

Electrolyzer part cost can be calculated as follows:

$$TPC_{EZ} = IC_{EZ} + OMC_{EZ} + RC_{EZ} \tag{7.13}$$

$$TPC_{EZ} = EZR * EZC + OMC_{EZ} + RC_{EZ} \tag{7.14}$$

where EZR: electrolyzer rate, kW. EZC: electrolyzer capital cost, \$/kW. OMC_{EZ}: operation and maintenance cost of electrolyzer, \$/kW. RC_{EZ}: electrolyzer replacement cost, \$/kW.

Ultracapacitor cost:

$$TPC_{UC} = IC_{UC} + RC_{UC} \tag{7.15}$$

$$TPC_{UC} = UCR * UCC + RC_{UC} \tag{7.16}$$

where UCR: ultracapacitor rate, kW. UCC: capital cost of UC, \$/kW. RC_{UC}: replacement cost of UC.

Power conditioner and wiring cost:

$$TPC_{pc\&w} = IC_{pc\&w} + OMC_{pc} + RC_{pc} \tag{7.17}$$

where $TPC_{pc\&w}$: total present cost of power conditioners and wiring, \$. $IC_{pc\&w}$: capital cost of power conditioners and wiring, \$/kW. OMC_{pc}: operation and maintenance cost of power conditioners, \$/kW. RC_{pc}: Power conditioners replacement cost, \$/kW.

Then total present cost can be calculated as:

$$TPC = TPC_{pv} + TPC_{WT} + TPC_{DRFC} + TPC_{UC} + TPC_{pc\&w} \tag{7.18}$$

Then substitute from Eqs. (7.4) and (7.18) in Eq. (7.3) for estimating the value of LEC, (\$/kWh)

7.4 Methodology and Constraints

The issue of optimization of the hybrid renewable energy system has been resolved by achieving the objective function taking into consideration all constraints, which includes the high reliability, the state of charge, minimum pollution, etc. This study focuses on the optimization of cost of energy and total present cost (TPC) taking into consideration many of constraints as the high reliability, amount of surplus and defect, the optimal sizing of the system and planning expansion for the future development. The LEC is developed as among the most effective economic profitability indicators of system economic analysis, which are considered as an objective function, the TPC and LEC can be calculated as [199]:

$$\min(\mathrm{TPC}) = \min\left(\mathrm{TPC}_{\mathrm{pv}} + \mathrm{TPC}_{\mathrm{WT}} + \mathrm{TPC}_{\mathrm{DRFC}} + \mathrm{TPC}_{\mathrm{UC}} + \mathrm{TPC}_{\mathrm{pc\&w}}\right) \quad (7.19)$$

$$\min(\mathrm{LEC}) = \min\left(\frac{\mathrm{TPC} \times \mathrm{CRF}}{\mathrm{LAE}}\right) \quad (7.20)$$

The system reliability was super reliability (the hybrid system satisfied the load by 100%).

7.4.1 PSO Optimization Algorithm

There are many techniques and algorithms that fall under the heading of artificial intelligence algorithms and techniques. Particle swarm algorithm is one of the most important of these algorithms; it is a modern technique based on randomly determined was presented in 1995 by Kennedy and Eberhart; it was applied with great success in the computational intelligence arena, solving global optimization problems and had exhibited many successful applications in many scientific fields. The PSO is using individual improvement called particles; these particles are moving during the problem area plus population cooperation and competition by following the current optimum particles; this technique is inspired by the social behavior of some animals, like fish schooling, the bird flocking, and the swarm theory. Each individual in PSO represents a possible solution assumed to have two properties [200]: velocity and position. These two variables are adjusted for each particle according to its own experience and the experience of its neighbors. Each particle wanders through the solution area and recalls the best objective function value (position), which has already been discovered; the fitness value is saved and known P_{best}. The PSO optimizer is seeking another best value which is acquired yet by any particle in the population. When a particle captures all the best population as its topological neighbors, the superior value is a global best, and it is called G_{best}.

Let the variable (x_i) refers to the position of particle (i) in the study space, and its speed is (v_i), so the (ith) from the particle can be represented as:

$$x_i = [x_{i1}, x_{i2}, x_{i3}, x_{i4}, \ldots x_{id}, \ldots x_{iN}] \tag{7.21}$$

In the N-dimensional space, each particle continuously registers the best solution that has reached so far during its flight (best fitness value P_{best}). The best past position of the ith particle is saved under the name vector and calculated by:

$$P_{\text{best}i} = [P_{\text{best}1}, P_{\text{best}2}, P_{\text{best}3}, P_{\text{best}4}, \ldots P_{\text{best}id}, \ldots P_{\text{best}iN}] \tag{7.22}$$

where $i = 1, 2, 3, \ldots, N$ is the number of particles in a swarm; the global best G_{best} refers to the best position which is ever achieved by all the individuals of the population and also tracked by the optimizer. The best particle of all the particles in the swarm is denoted by $G_{\text{best}d}$. The velocity of particle i is represented as:

$$v_i = [v_{i1}, v_{i2}, v_{i3}, v_{i4}, \ldots, v_{id}, \ldots, v_{iN}] \tag{7.23}$$

The velocity and position of each particle can be continuously adjusted based on the current velocity and the distance from $P_{\text{best}id}$ to $G_{\text{best}d}$:

$$v_i(t+1) = w(t)v_i(t) + c_1r_1(P_i(t) - X_i(t)) + c_2r_2(G(t) - X(t)) \tag{7.24}$$

$$X_i(t+1) = X_i(t) + \chi v_i(t+1) \tag{7.25}$$

In the Eqs. (7.24) and (7.25) the r_1 and r_2 are random real numbers drawn from [0, 1], c_1 and c_2 are acceleration constants that pull each particle towards P_{best} and G_{best} positions which are considered equal 1.0 in all applications. Thus, the particle flies through potential solutions towards $P_i(t)$ and $G(t)$ in a navigated way while still exploring new spaces by the random mechanism to escape from local optimum solution. There was no actual mechanism for controlling the velocity of a particle, so it was necessary to impose a maximum value $V_{\max}$, which controls the maximum travel distance in each iteration to avert this particle flying past good solutions. The χ is constriction factor which is used to limit velocity, here is $\chi = 0.7$. Also, after updating the positions, it must be checked that no particle violates the boundaries of search space. If a particle has violated the boundaries, it will be set at the boundary of the search area [200]. In Eq. (7.24), the inertia coefficient indicates the impact of the previous history of velocities on the current iteration one and is extremely very important to ensure convergent behavior. There are various strategies for calculating the inertia weight; these computations depend on the designer and can be supposed to be fixed or a variable [201]. In this study, the Particle Swarm Optimization technique (PSO) was developed to reach the best results by making the particles of the algorithm not only follow the traditional formula of the algorithm, in which the particles search and follow the nearest and best particle, but it was developed to

Table 7.1 PSO parameters

Parameters	Values
c_1 and c_2	1
r_1 and r_2	Randomly
w	Randomly
No. of iteration	100
No. of population	10

review all particles and look at the best particle in the group and follow it; the optimal solution was achieved in double speed, where it reached the optimal solution at a time rate better than 80% of conventional technology time and less than 20 repetitions only. In addition, this algorithm has been harnessed to achieve multiple objective functions. All the parameters of PSO algorithm are tabulated in Table 7.1; the inertia weight in this research is considered random and calculated by the equation:

$$w(t) = 0.5 + \frac{\text{rand}()}{2} \tag{7.26}$$

7.5 Results and Discussions

7.5.1 Analytic Methods

A new computer program has been proposed for carrying out the design that is discussed in Sects. 7.2 and 7.3. Hourly electrical load can be shown in Fig. 5.18 and optimum equipment's from previous chapters are used. Also the economic and technical parameters of the power system components can be shown in Table 7.2. The design power system age is 20 years, (LT = 20 years) and the interest rate is 0.02, ($r = 0.02$). The output of this program at different PELW and PELV is shown in Tables from 7.3, 7.4, 7.5, 7.6, 7.7, and 7.8). Also, results are shown in figures from 7.3, 7.4, 7.5, 7.6, 7.7, and 7.8.

Table 7.3 and Fig. 7.3 display ALC for HPS under different conditions of penetration levels using SPR-X22-360 solar panels with monthly best tilt angle and Gaia WTG type for Shark El-Oweinat site. From Table 7.3 and Fig. 7.4, the most economical case when PELW equals 100%, this means WT/RFC/UC is most economical HPS in this site, ALC equals 0.0932 $/kWh.

Table 7.4 and Fig. 7.4 display ALC for HPS under different conditions of penetration levels using SPR-X22-360 solar panels with fixed tilt angle equals latitude and Gaia WTG type for Shark El-Oweinat site. From these table and figure it can be concluded that the most economical case when PELW equals 100%, this means WT/RFC/UC is most economical HPS in this site, ALC equals 0.0932 $/kWh.

Table 7.2 The economic and technical parameters of the power system components [197, 202–204]

Item	IC ($/kW)	OMC ($/kW)	RC ($/kW)	T	N_{rep}
PV	1200	12	0	25	0
WTG	1988	60	0	20	0
FC part	400	500	400	10	2
EZ part	100	160	100	15	2
UC ($/kWh)	13	0	11	15	2
PC & Wiring	120	0	120	15	2

Table 7.3 HPS in Shark El-Oweinat site with monthly best tilt angle of solar panels and Gaia WTG type

Case	PELV (%)	PELW (%)	No. of modules	ONWTG	ALC $/kWh
1	100	0	52	0	0.1419
2	90	1	47	1	0.2084
3	80	20	42	1	0.1961
4	70	30	37	1	0.1839
5	60	40	32	1	0.1716
6	5	50	26	1	0.1569
7	40	60	21	1	0.1447
8	30	70	16	1	0.1324
9	20	80	11	1	0.1202
10	10	90	6	1	0.1079
11	0	100	0	1	0.0932

Table 7.4 HPS in Shark El-Oweinat site with fixed angle equal latitude of solar panels and Gaia WTG type

Case	PELV (%)	PELW (%)	No. of modules	ONWTG	ALC $/kWh
1	100	0	52	0	0.1107
2	90	10	47	1	0.1802
3	80	20	42	1	0.1710
4	70	30	37	1	0.1617
5	60	40	32	1	0.1524
6	50	50	26	1	0.1413
7	40	60	21	1	0.1321
8	30	70	16	1	0.1228
9	20	80	11	1	0.1136
10	10	9	6	1	0.1043
11	0	100	0	1	0.0932

Table 7.5 HPS in Marsa Alam Site with monthly best tilt angle of solar panels and Aircon WTG type

Case	PELV (%)	PELW (%)	No. of modules	ONWTG	ALC $/kWh
1	100	0	48	0	0.1321
2	90	10	44	1	0.1939
3	80	20	39	1	0.1816
4	70	30	34	1	0.1694
5	60	40	29	1	0.1571
6	50	50	24	1	0.1449
7	40	60	20	1	0.2067
8	30	70	15	2	0.1944
9	20	80	10	2	0.1822
10	10	90	5	2	0.1699
11	0	100	0	2	0.1576

Table 7.6 HPS in Marsa Alam with fixed angle equals latitude of solar panels and Aircon WTG type

Case	PELV (%)	PELW (%)	No. of modules	ONWTG	ALC $/kWh
1	100	0	52	0	0.1107
2	90	10	47	1	0.1731
3	80	20	42	1	0.1638
4	70	30	37	1	0.1545
5	60	40	32	1	0.1453
6	50	50	26	1	0.1342
7	40	60	21	1	0.1965
8	30	70	16	2	0.1873
9	20	80	11	2	0.1780
10	10	90	6	2	0.1688
11	0	100	0	2	0.1576

Table 7.7 HPS in Minia with monthly best tilt angle of solar panels and Aircon WTG type

Case	PELV (%)	PELW (%)	No. of modules	ONWTG	ALC $/kWh
1	100	0	56	0	0.1517
2	90	10	51	1	0.2110
3	80	20	45	1	0.1963
4	70	30	40	2	0.2557
5	6	40	34	2	0.2410
6	50	5	28	2	0.2263
7	40	60	23	3	0.2856
8	30	70	17	3	0.2709
9	20	80	12	3	0.3302
10	10	90	6	4	0.3155
11	0	100	0	4	0.3008

Table 7.8 HPS in Minia with fixed angle equals latitude of solar panels and Aircon WTG type

Case	PELV (%)	PELW (%)	No. of modules	ONWTG	ALC $/kWh
1	100	0	56	0	0.1181
2	90	10	51	1	0.1805
3	80	20	45	1	0.1694
4	70	30	40	2	0.2317
5	60	40	34	2	0.2206
6	50	50	28	2	0.2095
7	40	60	23	3	0.2718
8	30	70	17	3	0.2607
9	20	80	12	3	0.3230
10	10	90	6	4	0.3119
11	0	100	0	4	0.3008

Table 7.5 and Fig. 7.5 display ALC for HPS under different conditions of penetration levels using SPR-X22-360 solar panels with monthly best tilt angle and Aircon WTG type for Marsa Alam Site. From these table and figure it can be concluded that the most economical case when PELV equals 100%, this means PV/RFC/UC is most economical HPS in this site, ALC equals 0.1321 $/kWh.

Table 7.6 and Fig. 7.6 display ALC for HPS under different conditions of penetration levels using SPR-X22-360 solar panels with fixed tilt angle equals latitude and Aircon WTG type for Marsa Alam Site. From these table and figure it can be concluded that the most economical case when PELV equals 100%, this means PV/RFC/UC is most economical HPS in this site, ALC equals 0.1107 $/kWh.

Table 7.7 and Fig. 7.7 display ALC for HPS under different conditions of penetration levels using SPR-X22-360 solar panels with monthly best tilt angle and Aircon WTG type for Minia. From these table and figure it can be concluded that the most economical case when PELV equals 100%, this means PV/RFC/UC is most economical HPS in this site, ALC equals 0.1517 $/kWh.

Table 7.8 and Fig. 7.8 display ALC for HPS under different conditions of penetration levels using SPR-X22-360 solar panels with fixed tilt angle equals latitude and Aircon WTG type for Minia. From these table and figure it can be concluded that the most economical case when PELV equals 100%, this means PV/RFC/UC is most economical HPS in this site, ALC equals 0.1181 $/kWh (Fig. 7.9).

7.5.2 PSO Results and Discussion

This study focuses on the economic optimization of hybrid renewable systems using Particle Swarm Optimization (PSO), with an objective function based on minimizing the total present cost (TPC) and the levelized energy cost (LEC) of the hybrid renewable system consisting of PV panels/wind turbine/regenerative fuel cells/

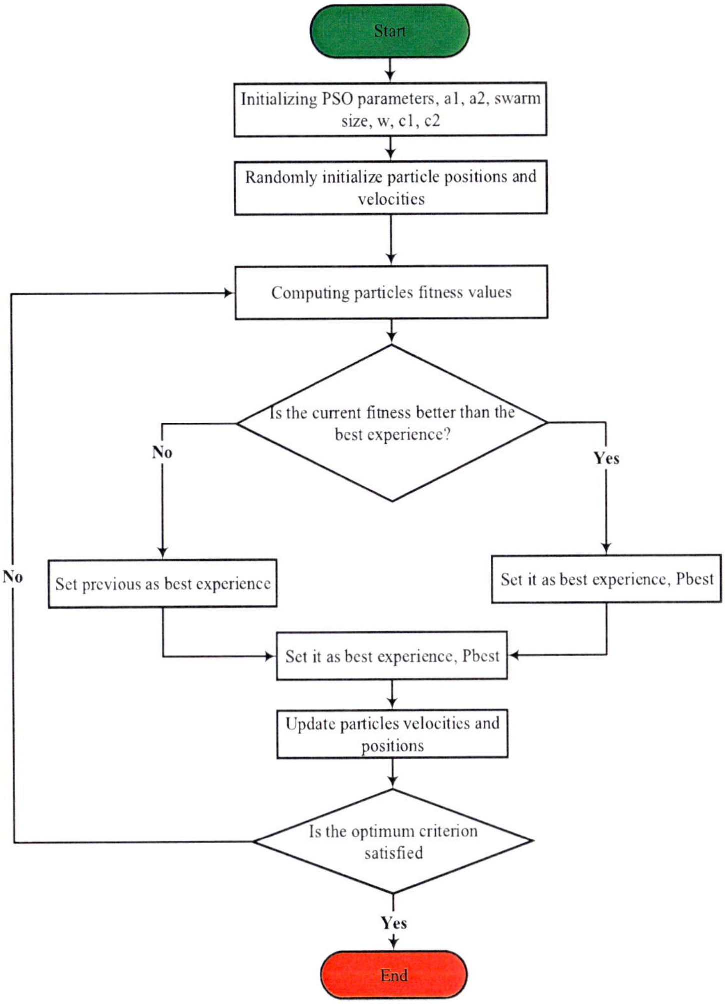

Fig. 7.3 The process of the PSO algorithm

ultracapacitor, taking into account many of the constraints, like the meet of the electrical demand with high reliability, equality of surplus/deficit energy, the optimal sizing of the system and planning expansion for the future development. The PSO algorithm program was developed using the MATLAB environment. All variables and data concerned the hybrid system and renewable energy sources at the location are inserted, as solar insolation data, wind speeds, temperature data required, regenerative fuel cell specs, ultracapacitors specs, all prices details as the total cost, maintenance and operating costs, the coordinates of the location, the life of project, the nearness of the national electric grid, the numbers and prices of all components of hybrid power system generation if they already exist before the creation of the

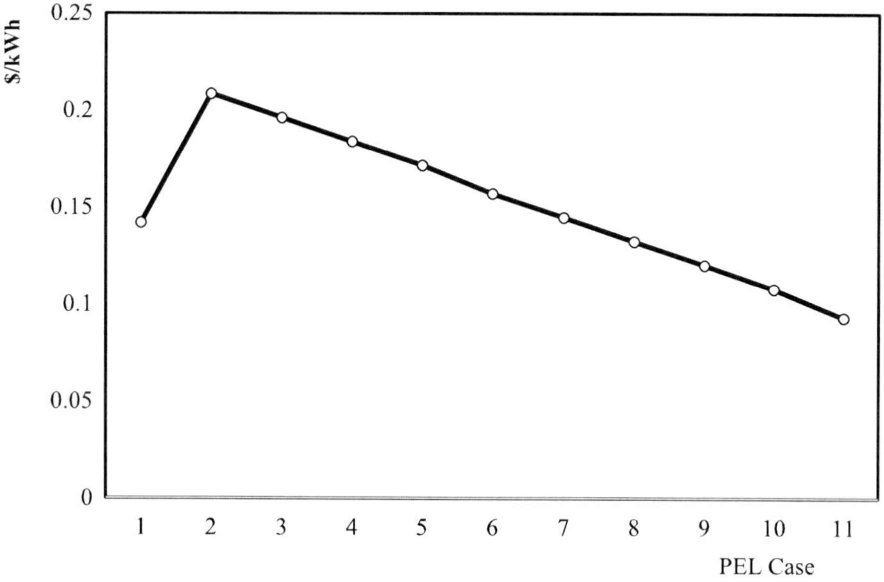

Fig. 7.4 ALC of HPS in Shark El-Oweinat site with monthly best tilt angle of solar panels and Gaia WTG type

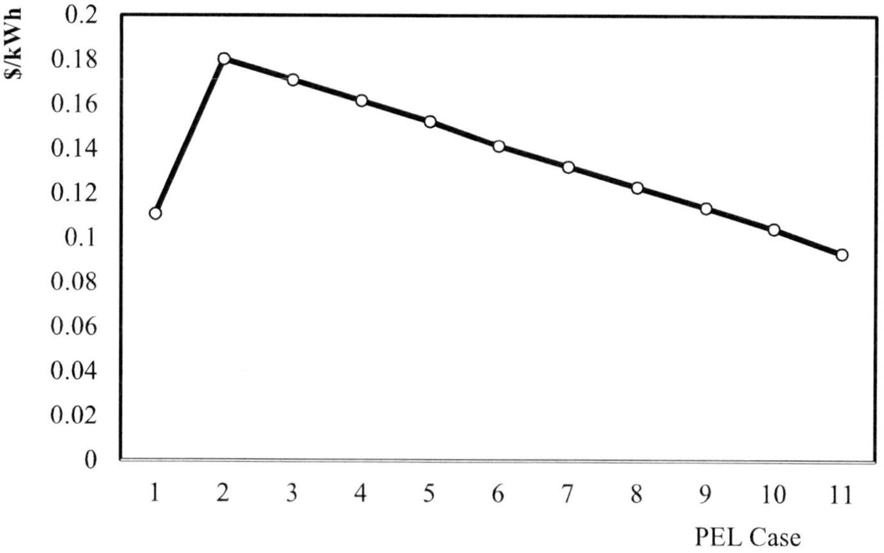

Fig. 7.5 ALC of HPS in Shark El-Oweinat site with fixed angle equal latitude of solar panels and Gaia WTG type

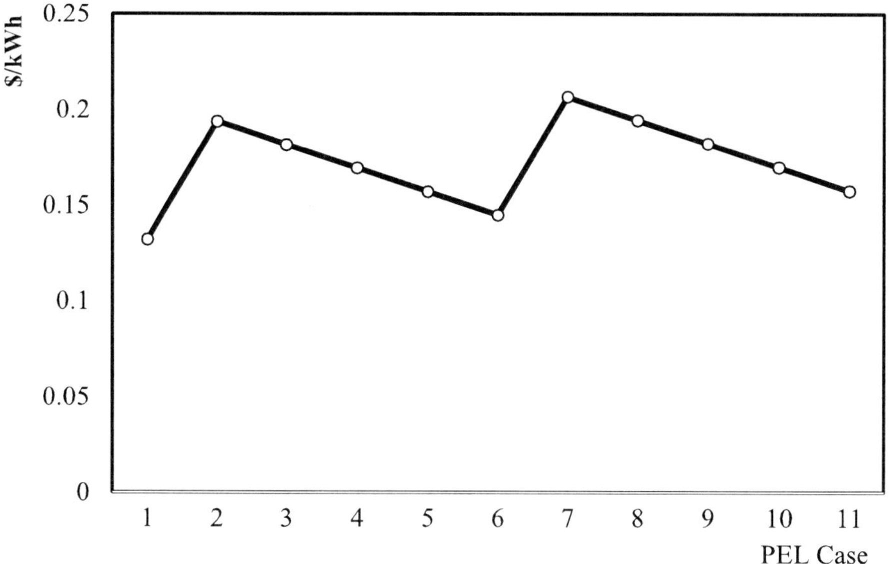

Fig. 7.6 ALC of HPS in Marsa Alam Site with monthly best tilt angle of solar panels and Aircon WTG type

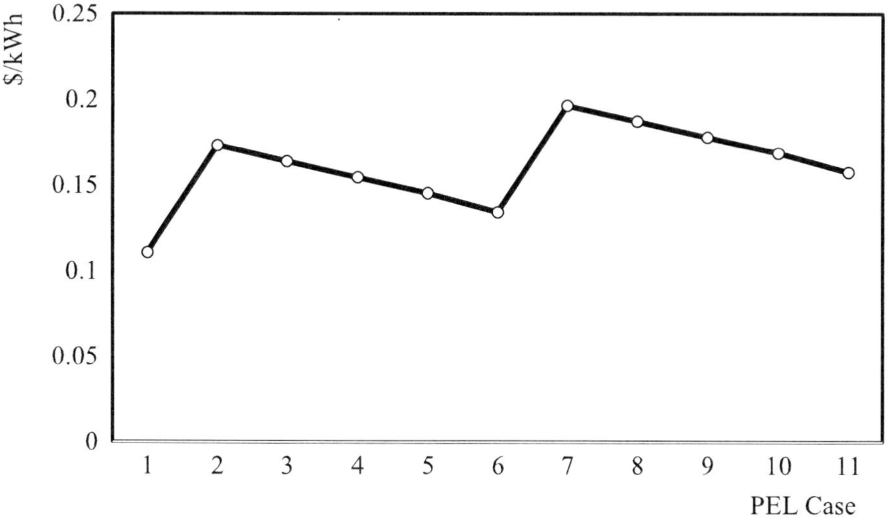

Fig. 7.7 ALC of HPS in Marsa Alam with fixed angle equal latitude of solar panels and Aircon WTG type

Fig. 7.8 ALC of HPS in Minia with monthly best tilt angle of solar panels and Aircon WTG type

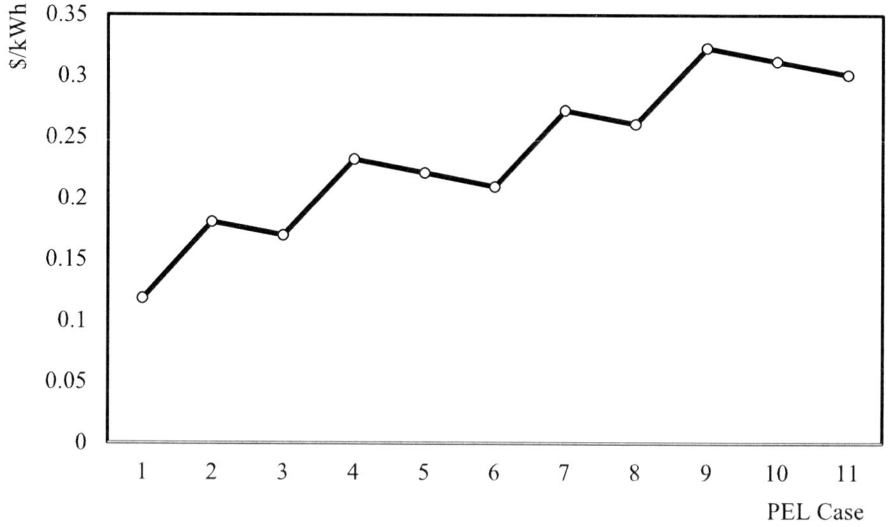

Fig. 7.9 ALC of HPS in Minia with fixed angle equal latitude of solar panels and Aircon WTG type

Table 7.9 Optimum PEL and LEC for Marsa Alam

Best tilt angle			Latitude		
PELV	PELW	LEC	PELV	PELW	LEC
100%	0%	0.1323	100%	0%	0.1109

Table 7.10 Optimum PEL and LEC for Shark El-Oweinat

Best tilt angle			Latitude		
PELV	PELW	LEC	PELV	PELW	LEC
0%	100%	0.0934	0%	100%	0.0934

Table 7.11 Optimum PEL and LEC for Minia

Best tilt angle			Latitude		
PELV	PELW	LEC	PELV	PELW	LEC
100%	0%	0.1519	100%	0%	0.1183

Table 7.12 Comparison between ALC results for analytical method and PSO method

Location	Solar panels tilt	PELV (%)	PELW (%)	ALC analy. ($/kWh)	ALC (PSO) ($/kWh)
Marsa Alam	Month. best	100	0	0.1321	0.1323
	Latitude	100	0	0.1107	0.1109
Shark El-Oweinat	Month. best	0	100	0.0932	0.0934
	Latitude	0	100	0.0932	0.0934
Minia	Month. best	100	0	0.1517	0.1519
	Latitude	100	0	0.1181	0.1183

project, etc. The particle swarm optimization design is to get highly accurate solutions with the minimum time.

The output results from PSO programs can be shown as follows (Tables 7.9, 7.10, and 7.11):

Comparison is done between ALC results from analytical method and PSO method. Table 7.12 and Fig. 7.10 display this comparison.

It can be noted from Table 7.12 and Fig. 7.10 as follows:

1. ALC results for analytical method and PSO method are approximately the same.
2. In Marsa Alam location using PVPS with fixed angle of solar panels, equal latitude is economical with ALC = 0.1109.
3. Wind Energy system only is the economic in Shark El-Oweinat location with ALC = 0.0934.
4. In Minia location using PVPS with fixed angle of solar panels, equal latitude is economical with ALC = 0.1183.
5. Hybrid energy system is not economical in all chosen loacations.

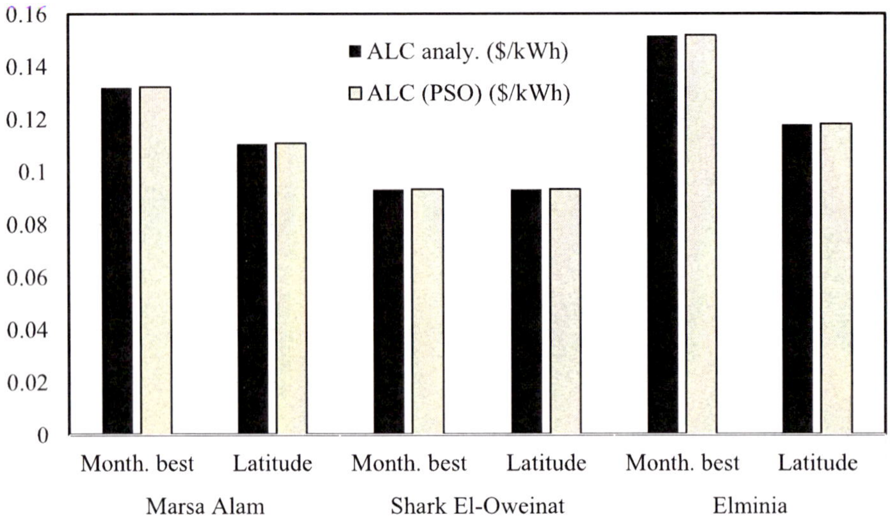

Fig. 7.10 Comparison between ALC results for analytical method and PSO method

7.6 Summary

This chapter proposes a powerful system of hybrid (solar panels, wind turbines and regenerative fuel cells and ultracapacitor). The economic study of the hybrid system is in two ways: analytical method and artificial intelligence method; it has appeared that the final results of the two methods are almost identical. The results showed that there are locations where solar panels are more economically used than air turbines and vice versa. in order to small electrical loads, the mixed system of solar panels and wind turbines is costly.

Chapter 8
Conclusions and Recommendations for Future Work

8.1 Discussions and Conclusions

This book presents the optimum design and sizing of the renewable energy system to supply communication stations. Now, there are a lot of failures in the power supply system of communication stations especially in remote and distant locations, so this book determined the failure points and tried to treat the failure points.

The following are the salient discussions and conclusions that can be drawn from this book:

1. Batteries in the power system of traditional communication stations are replaced by ultracapacitor, which represents the battery of the future.
2. Diesel machine in the power system of traditional communication stations is replaced by regenerative fuel cells, which do without fuel and water source. It is environment-friendly.
3. Base station power consumption equals 80% of rated at low traffic, 90% at average traffic and 100% of rated at high traffic.
4. Regenerative fuel cells is one of the optimum solutions to overcome water shortage when the hydrogen is needed.
5. The efficiency of DRFC is greater than the efficacy of URFC.
6. PEM FC is the more suitable FC for communication application, in order to low operating temperature.
7. BCAP2000 UC type is the optimum choice; it can supply the full load for time equal to 28.694 s at simple sizing and 26.73 at accurate sizing.
8. There is a difference between UC simple sizing, and UC accurate sizing equals 7.35% at the optimal choice.
9. DC-DC converters are four types, and they used to convert a DC voltage from the solar array into a different value of a DC output voltage. Regulate the DC voltage from the solar array depending on the load and line variations and other functions.

A. A. Elbaset, S. Ata, *Hybrid Renewable Energy Systems for Remote Telecommunication Stations*, https://doi.org/10.1007/978-3-030-66344-5_8

10. There are four solar panels type used to determine the optimum, the selected solar panels are: FVG 185M-MC, YS280-60, RS-M300, and SPR-X22-360
11. There is an hourly profile load for a communication station that is used to determine optimum solar cells area.
12. The design is applied to three random locations in Egypt (Marsa Alam, Minia, and Shark El Oweinat).
13. OSCA is inversely proportional to PV module efficiency (high PV module efficiency means decreased OSCA and vice versa).
14. The monthly best tilt angle gives the lower OSCA from all selected PV modules. Related with latitude angle.
15. For lower PV module efficiency the percentage differences in OSCA are low then as PV module efficacy is increased, the difference in OSCA increases.
16. SPR-X22-360 solar panel is optimum for all location because it gives minimum OSCA required in cases of latitude tilt angle and monthly optimum tilt angle for the selected sites.
17. Eleven wind turbines are used to obtain the optimum wind turbine. These wind turbines are Aeolos-V, Iskra, Sirocco, Aircon, Gaia, Jonica Impianti, Eoltec, Fuhrländer (FL), Pitchwind (H20), Pitchwind (H62), and FL 100.
18. Gaia wind turbine is the optimum for Shark El-Oweinat location because it gives the highest capacity factor that equals 0.3413.
19. An Aircon wind turbine is the optimum for Marsa Alam location because it gives the highest capacity factor, CF, that equals 0.1729; this value of CF is small, so this location is poor in wind energy.
20. An Aircon wind turbine is optimum for the Shark Minia location because it gives the highest capacity factor, CF, that equals 0.0712. This value of CF is small, so this location is poor in wind energy.
21. The optimum number of wind turbine in Shark El-Oweinat equals one wind turbine of Gaia type. It provides annually 32,325.8293 kWh output energy.
22. The optimum number of wind turbine in Minia equals 4 wind turbines of Aircon type. It produces annually 24,808.5077 kWh output energy.
23. The optimum number of wind turbine in Marsa Alam site equals 2 wind turbines of Gaia type. It produces annually 30,535.5135 kWh output energy.
24. Optimum levelized energy cost. ALC for Marsa Alam is equal to 0.1109 $/kWh and it found at a penetration level of PV equal to 100% at a fixed tilt angle of solar panels at the same latitude.
25. Optimum levelized energy cost. ALC for Shark El-Oweinat is equal to 0.0934 $/kWh and it found at a penetration level of WES equal to 100%.
26. Optimum levelized energy cost. ALC for Minia is equal to 0.1183 $/kWh and it found at a penetration level of PV equal to 100% at a fixed tilt angle of solar panels at the same latitude.

References

1. S. Paudel, et al., Technical and economic assessment of renewable energy sources for telecom application: a case study of Nepal Telecom, in *Proceedings of the 5th International Conference on Power and Energy Systems*, Kathmandu, Nepal, 2013, pp. 28–30
2. M. Yeshalem, B. Khan, Design of an off-grid hybrid PV/wind power system for remote mobile base station: A case study. AIMS Energy **5**(1), 96–112 (2017)
3. A. Kumar et al., 3GPP LTE: The future of mobile broadband. Wireless Pers. Commun. **62**(3), 671–686 (2012)
4. W. Zhou et al., Current status of research on optimum sizing of stand-alone hybrid solar–wind power generation systems. Appl. Energy **87**(2), 380–389 (2010)
5. O. Ekren, B. Ekren, Size optimization of a PV/wind hybrid energy conversion system with battery storage using simulated annealing. Appl. Energy **87**(2), 592–598 (2010)
6. L. Chiaraviglio, et al., Energy-aware UMTS access networks, in *The 11th International Symposium on Wireless Personal Multimedia Communications (WPMC'08)*, 2008.
7. N. El Bassam, P. Maegaard, M. Schlichting, *Distributed Renewable Energies for Off-Grid Communities: Strategies and Technologies Toward Achieving Sustainability in Energy Generation and Supply* (Newnes, Boston, 2013)
8. S. Roy, G. Marketing, *Energy Logic for Telecommunications, White Paper for Emerson Network Power* (Citeseer, 2008)
9. M.M. Rahman, *Overview of Energy Saving Aspects in 2G and 3G Mobile Communication Networks*, Master thesis, University of Gavle, Stockholm, 2009
10. G. Van Der Zanden, *The Smart Grid in Europe: The Impact of Consumer Engagement on the Value of the European Smart Grid*, Master thesis, Lund University, Sweden, 2011
11. L. Ardito et al., Smart grid technologies in Europe: An overview. Energies **6**(1), 251–281 (2013)
12. F. Rotaru et al., Two-stage distributed generation optimal sizing with clustering-based node selection. Int. J. Electr. Power Energy Syst. **40**(1), 120–129 (2012)
13. I. Dincer, Renewable energy and sustainable development: a crucial review. Renew. Sustain. Energy Rev. **4**(2), 157–175 (2000)
14. O. Ellabban, H. Abu-Rub, F. Blaabjerg, Renewable energy resources: Current status, future prospects and their enabling technology. Renew. Sustain. Energy Rev. **39**, 748–764 (2014)
15. N.S. Hasan et al., Review of storage schemes for wind energy systems. Renew Sustain. Energy Rev. **21**, 237–247 (2013)
16. B. Robyns et al., *Electricity Production from Renewable Energies* (Wiley, Chichester, 2012)

A. A. Elbaset, S. Ata, *Hybrid Renewable Energy Systems for Remote Telecommunication Stations*, https://doi.org/10.1007/978-3-030-66344-5

17. D. Kaundinya et al., Grid-connected versus stand-alone energy systems for decentralized power—A review of literature. Renew. Sustain. Energy Rev. **13**(8), 2041–2050 (2009)
18. I. López et al., Review of wave energy technologies and the necessary power-equipment. Renew. Sustain. Energy Rev. **27**, 413–434 (2013)
19. I. Fridleifsson, et al., The possible role and contribution of geothermal energy to the mitigation of climate change, in *IPCC Scoping Meeting on Renewable Energy Sources, Proceedings*, Luebeck, Germany, vol. 20, no. 25, 2008, pp. 59–80.
20. S. Rehman et al., Xylitol: A review on bioproduction, application, health benefits, and related safety issues. J. Crit. Rev. Food Sci. Nutr. **55**(11), 1514–1528 (2015)
21. S. Ruin, G. Sidén, *Small-Scale Renewable Energy Systems: Independent Electricity for Community, Business and Home* (CRC Press, Boca Raton, 2019)
22. G. Jansen, et al., Validation of autonomous renewable energy hybrid wind/photovoltaic/RHFC prototype for the cell tower industry using MATLAB/Simulink, in *9th International Conference on Advanced Materials*, 2019
23. R. Kaur, V. Krishnasamy, N. Kandasamy, Optimal sizing of wind–PV-based DC microgrid for telecom power supply in remote areas. IET Renew. Power Gen. **12**(7), 859–866 (2018)
24. M. Rahman, *Resilient Hybrid Energy System (RHES) for powering cellular base station during natural disasters*, Master thesis, Lamar University-Beaumont, 2018
25. B.M. Altinoz, *Grid Integration of Combined Solar-Wind-Batteryenergy Systems with Economic Analysis*, Master thesis, University of Massachusetts Lowell, 2017
26. S. Cordiner et al., Fuel cell based Hybrid Renewable Energy Systems for off-grid telecom stations: Data analysis from on field demonstration tests. Appl. Energy **192**(15), 508–518 (2017)
27. A.A. Elbaset, M.S. Hassan, *Design and Power Quality Improvement of Photovoltaic Power System(eBook)* (Springer, New York, 2017)
28. S. Belkhiri, A. Chaker, Optimization of hybrid PV/wind system for remote telecom station, a case study of different sites in Algeria, in *Biological, and Environmental Engineering*, Vietnam, 2016, pp. 23–25
29. S. Devabakthuni, *Grid-Connected Wind and Photovoltaic System*, Master thesis, Department of Electrical Engineering Northern Illinois University, 2016
30. D. Shen, *Hybrid Wind-Solar-Storage Energy Harvesting Systems*, Master thesis, Purdue University, 2016.
31. W. Margaret Amutha, V. Rajini, Techno-economic evaluation of various hybrid power systems for rural telecom. Renew. Sustain. Energy Rev. **43**, 553–561 (2015)
32. D. Sharma, R.S. Meena, D. Birla, PV-wind system with fuel cell & electrolyzer. Int. J. Eng. Res. **4**(12), 673–679 (2015)
33. T. Salih, Y. Wang, M. Adam, Renewable micro hybrid system of solar panel and wind turbine for telecommunication equipment in remote areas in Sudan. Energy Proc. **61**, 80–83 (2014)
34. E.O. Noguera, *Comparative LCA of Stand-Alone Power Systems Applied To Remote Cell Towers*, Master thesis, Purdue University, 2014
35. L.J. Olatomiwa, S. Mekhilef, A.S.N. Huda, Optimal sizing of hybrid energy system for a remote telecom tower: A case study in Nigeria, in *2014 IEEE Conference on Energy Conversion (CENCON)*, 2014, pp. 243–247
36. K. Kusakana, H.J. Vermaak, Hybrid renewable power systems for mobile telephony base stations in developing countries. Renew Energy **51**, 419–425 (2013)
37. E.M. Natsheh, *Hybrid Power Systems Energy Management Based on Artificial Intelligence*, PhD thesis, Manchester Metropolitan University, 2013
38. H.A.H. Hassan, L. Nuaymi, A. Pelov, Renewable energy in cellular networks: A survey, in *2013 IEEE Online Conference on Green Communications (OnlineGreenComm)*, 2013, pp. 1–7
39. Q. Zhai, *Application Potential of Solar PV, Wind and Fuel Cells Technologies for Global Sustainability Improvement*, PhD thesis, University of Wisconsin-Milwaukee, 2012

40. A.M. Al-Sharafi, *Modeling and Analysis of Hybrid PV/Wind Off-Grid Power Generation System in the Kingdom of Saudi Arabia*, Master thesis, King Fahd University of Petroleum and Minerals (Saudi Arabia), 2011
41. S. Paudel, et al., Optimization of hybrid PV/wind power system for remote telecom station, in *2011 International Conference on Power and Energy Systems*, 2011, pp. 1–6
42. P. Nema, R. Nema, S. Rangnekar, PV-solar/wind hybrid energy system for GSM/CDMA type mobile telephony base station. Int. J. Energy Environ. **1**(2), 359–366 (2010)
43. A.A. Elbaset, *Study of Interconnecting Issues of Photovoltaic/Wind Hybrid System with Electric Utility Using Artificial Intelligence*, PhD thesis, Faculty of Engineering, Minia University, 2006
44. C. Lubritto et al., Telecommunication power systems: energy saving, renewable sources and environmental monitoring, in *INTELEC 2008-IEEE 30th International Telecommunications Energy Conference*, pp. 1–4
45. Vereecken et al., Power consumption in telecommunication networks: overview and reduction strategies. IEEE Commun. Mag. **49**(6), 62–69 (2011)
46. C. Han et al., Green radio: radio techniques to enable energy-efficient wireless networks. IEEE Commun. Mag. **49**(6), 46–54 (2011)
47. P. Ziridis, *Game theoretic analysis of sharing telecommunication infrastructure*, Master thesis, International Hellenic University, Greece, 2015
48. M. Marsan, et al., Optimal energy savings in cellular access networks, in *IEEE International Conference on Communications Workshops*, 2009, pp. 1–5
49. M. Deruyck et al., Reducing the power consumption in wireless access networks: Overview and recommendations. Prog. Electromagn. Res. Pier **132**, 255–274 (2012)
50. O. Arnold, et al., Power consumption modeling of different base station types in heterogeneous cellular networks, in *2010 IEEE Future Network & Mobile Summit*, 2010, pp. 1–8
51. M. Deruyck et al., Comparison of power consumption of mobile WiMAX, HSPA and LTE access networks, in *9th IEEE Conference of Telecommunication, Media and Internet*, 2010, pp. 1–7
52. J. Lorincz, T. Garma, G. Petrovic, Measurements and modelling of base station power consumption under real traffic loads. Sensors **12**(4), 4281–4310 (2012)
53. M.P.D. Chilukuri, *Power Profiling of Network Switches*, Master thesis, Faculty of Computing Blekinge Institute of Technology, Sweden, 2017
54. Y. Zhang, et al., Energy efficiency in telecom optical networks, in *IEEE Communications Surveys & Tutorials*, vol. 12, no. 4, 2010, pp. 441–458
55. E. Hossain, K. Leung, *Wireless Mesh Networks Architectures and Protocols* (Springer, Berlin, 2008)
56. S. Thota et al., Computing for Rural Empowerment: Enabled by Last-Mile Telecommunications (Extended Version), Technical Report (2013)
57. R. Lu et al., GRS: The green, reliability, and security of emerging machine to machine communications. IEEE Commun. Mag. **49**(4), 28–35 (2011)
58. R. Bolla et al., The potential impact of green technologies in next-generation wireline networks: Is there room for energy saving optimization. IEEE Commun. Mag. **49**(8), 80–86 (2011)
59. G. Kramer et al., Evolution of optical access networks: Architectures and capacity upgrades. Proc. IEEE **100**(5), 1188–1196 (2012)
60. S. Bindhaiq et al., Recent development on time and wavelength-division multiplexed passive optical network (TWDM-PON) for next-generation passive optical network stage 2 (NG-PON2). Opt. Switch. Netw. **15**, 53–66 (2015)
61. C. Lin, *Broadband Optical Access Networks and Fiber-to-the-Home* (Wiley Online Library, 2006)
62. S. Aleksić, A. Lovrić, Power consumption of wired access network technologies, in *2010 7th International Symposium on Communication Systems, Networks & Digital Signal Processing (CSNDSP 2010)*, IEEE, 2010, pp. 147–151

63. K. Wang et al., Migration strategies for FTTx solutions based on active optical networks. IEEE Commun. Mag. **54**(2), 78–85 (2016)
64. B. Bidoggia, et al., Degign of a fuel cell based backup system for telecom applications, in *INTELEC 2008, IEEE 30th International Telecom Energy Conference*, IEEE, 2008
65. O. Sharaf, M. Orhan, An overview of fuel cell technology: Fundamentals and applications. *Renew. Sustain. Energy Rev.* **32**, 810–853 (2014)
66. A. Abd-Elbaset et al., Optimal design of a PV/fuel cell hybrid power system for the city of Brest in France, in *2014 First International Conference on Green Energy ICGE 2014*, IEEE, pp. 119–123
67. B. Gou, W. Na, B. Diong, *Fuel Cells: Modeling, Control, and Applications* (CRC Press, Boca Raton, 2017)
68. W. Carter, B. Diong, Model of a regenerative fuel cell-supported wind turbine ac power generating system, in *39th IAS Annual Meeting Conference Record of the 2004 IEEE Industry Applications Conference*, vol. 4, IEEE, 2004, pp. 2778–2785
69. A. Alaswad et al., Developments in fuel cell technologies in the transport sector. Int. J. Hydrogen Energy **41**(37), 16499–16508 (2016)
70. A. Dicks, D. Rand, *Fuel Cell Systems Explained* (Wiley Online Library, 2018)
71. F. Zhang et al., The survey of key technologies in hydrogen energy storage. Int. J. Hydrogen Energy **41**(33), 14535–14552 (2016)
72. S. Singhal, Solid oxide fuel cells for stationary, mobile, and military applications. Solid State Ion **152**, 405–410 (2002)
73. B. Kumaragurubaran et al., Recent development in fuel cell as an alternate fuel. Mater. Sci. Res. India **5**(2), 339–346 (2008)
74. T. Wilberforce et al., Advances in stationary and portable fuel cell applications. Int. J. Hydrogen Energy **41**(37), 16509–16522 (2016)
75. X. Zhang et al., Towards a smart energy network: The roles of fuel/electrolysis cells and technological perspectives. Int. J. Hydrogen Energy **40**(21), 6866–6919 (2015)
76. M. Cifrain, K. Kordesch, *Hydrogen/Oxygen (Air) Fuel Cells with Alkaline Electrolytes* (Wiley Online Library, 2010)
77. M. Lourens, H. Pienaar, Design and development of a zinc-air fuel cell for telecommunication systems, in *Proceedings*, vol. 1, 2006, pp. 1–5
78. X.M. Ge et al., Solid oxide fuel cell anode materials for direct hydrocarbon utilization. Adv. Energy Mater. **2**(10), 1156–1181 (2012)
79. T. Kousksou et al., Energy storage: Applications and challenges. Solar Energy Mater. Solar Cells **120**, 59–80 (2014)
80. E. Testa, *Multidisciplinary Design and Optimization of Innovative Electrical Power Systems for Aerospace Applications*, PhD thesis, Politecnico di Torino, 2013
81. A. Doddathimmaiah, *Unitised Regenerative Fuel Cells in Solar-Hydrogen Systems for Remote Area Power Supply*, PhD thesis, RMIT University, 2008
82. B. Paul, J. Andrews, PEM unitised reversible/regenerative hydrogen fuel cell systems: State of the art and technical challenges. Renew. Sustain. Energy Rev. **79**, 585–599 (2017)
83. J. Andrews, A. Doddathimmaiah, Regenerative fuel cells, in *Materials for Fuel Cells*, (Elsevier, Amsterdam, 2008), pp. 344–385
84. S.I.H. Ata, H.H. El-Tamaly, Optimum design of photovoltaic regenerative fuel cell power system for a remote telecom station. Int. Res. J. Eng. Technol. **5**(3), 1958–1964 (2018)
85. K. Burke, High energy density regenerative fuel cell systems for terrestrial applications. IEEE Aerospace Electron. Syst. Mag. **14**(12), 23–34 (1999)
86. S. Grigoriev et al., Pure hydrogen production by PEM electrolysis for hydrogen energy. Int. J. Hydrogen Energy **31**(2), 171–175 (2006)
87. S. Dutta, E. Chemistry, A review on production, storage of hydrogen and its utilization as an energy resource. J. Ind. Eng. Chem. **20**(4), 1148–1156 (2014)

88. S. Satyapal et al., The US Department of Energy's National Hydrogen Storage Project: Progress towards meeting hydrogen-powered vehicle requirements. Catal. Today **120**(3–4), 246–256 (2007)
89. M.M. Elamari, Optimisation of photovoltaic-powered electrolysis for hydrogen production for a remote area in libya, PhD thesis, The University of Manchester (United Kingdom), 2011
90. K. Hirose, *Handbook of Hydrogen Storage: New Materials for Future Energy Storage* (Wiley, Hoboken, 2010)
91. M. Sankir, N. Sankir, J. Wiley, *Hydrogen Production Technologies* (Wiley Online Library, 2017)
92. R.B. Gupta, *Hydrogen Fuel: Production, Transport, and Storage* (CRC Press, Boca Raton, 2008)
93. L. Gandia, G. Arzamedi, P. Diéguez, *Renewable Hydrogen Technologies: Production, Purification, Storage, Applications and Safety* (Elsevier, Amsterdam, 2013)
94. A. Cultura, Z. Salameh, Modeling, evaluation and simulation of a supercapacitor module for energy storage application, in *International Conference on Computer Information Systems and Industrial Applications*, (Atlantis Press, Paris, 2015)
95. D. Petreus et al., Modeling and sizing of supercapacitors. Adv. Electr. Comput. Eng. **8**(2), 15–22 (2008)
96. A. Abd-Elbaset, Smoothing of grid-connected wind-diesel power output using energy capacitor system. J. Eng. Sci. Technol. Rev. **7**(2), 47–52 (2014)
97. L. Seim, *Modeling, Control and Experimental Testing of a Supercapacitor/Battery Hybrid System: Passive and Semi-active Topologie*, Master thesis, Norwegian University of Life Sciences, 2012
98. A. Shukla et al., Electrochemical supercapacitors: Energy storage beyond batteries. Curr. Sci. **79**(12), 1656–1661 (2000)
99. A. Yu, V. Chabot, J. Zhang, *Electrochemical Supercapacitors for Energy Storage and Delivery: Fundamentals and Applications* (CRC Press, Boca Raton, 2013)
100. L. Zhang et al., Experimental impedance investigation of an ultracapacitor at different conditions for electric vehicle applications. J. Power Sources **287**, 129–138 (2015)
101. V. Shah et al., Improved method for characterization of ultracapacitor by constant current charging. Int. J. Model. Optim. **2**(3), 290–294 (2012)
102. J. Jamnik, J. Maier, Generalised equivalent circuits for mass and charge transport: chemical capacitance and its implications. Phys. Chem. Chem. Phys. **3**(9), 1668–1678 (2001)
103. D. Caetano et al., Differential capacitance of an electric double layer with asymmetric solvent-mediated interactions: mean-field theory and Monte Carlo simulations. Phys. Chem. Chem. Phys. **19**(35), 23971–23981 (2017)
104. C.W. Miller, *Temperature Characterization of the Ultracapacitor Serial Resistance Using a Constant Voltage Source*, Master's thesis, University of Tennessee, 2008
105. S. Ahmad Hamidi et al., Modeling and management of batteries and ultracapacitors for renewable energy support in electric power systems–an overview. Electr. Power Components Syst. **43**(12), 1434–1452 (2015)
106. L. Shi, M. Crow, Comparison of ultracapacitor electric circuit models, in 2008 *IEEE Power and Energy Society General Meeting-Conversion and Delivery of Electrical Energy in the 21st Century*, IEEE, 2008, pp. 1–6
107. R. Dougal, L. Gao, S. Liu, Ultracapacitor model with automatic order selection and capacity scaling for dynamic system simulation. J. Power Sources **126**(1–2), 250–257 (2004)
108. Md. Shadiul Islam et al. Modiling of a Double-layer Capacitor with individual branch response, in *WCECS*, 20–22 Oct 2010
109. E. Frackowiak, F. Béguin, *Supercapacitors: Materials, Systems and Applications* (Wiley, Weinheim, 2013)
110. G. Wang, L. Zhang, A review of electrode materials for electrochemical supercapacitors. Chem. Soc. Rev. **41**(2), 797–828 (2012)
111. J. Schonberger, *Modeling a Supercapacitor using PLECS*, vol 3 (Plexim GmbH, Zurich, 2010)

112. O. Bohlen, J. Kowal, D. Sauer, Ageing behaviour of electrochemical double layer capacitors: Part I. Experimental study and ageing model. J. Power Sources **172**(1), 468–475 (2007)
113. R. Faranda, A new parameters identification procedure for simplified double layer capacitor two-branch model. Electr. Power Syst. Res. **80**(4), 363–371 (2010)
114. A. Nadeau, G. Sharma, T. Soyata, State-of-charge estimation for supercapacitors: A kalman filtering formulation, in *2014 IEEE International Conference on Acoustics, Speech and Signal Processing (ICASSP)*, 2014, pp. 2194–2198
115. J.F. Pedrayes et al., Optimization of supercapacitor sizing for high-fluctuating power applications by means of an internal-voltage-based method. *Energy* **183**, 504–513 (2019)
116. Z. Sen, *Solar Energy Fundamentals and Modeling Techniques: Atmosphere, Environment, Climate Change and Renewable Energy* (Springer, Berlin, 2008)
117. D. Rekioua, E. Matagne, *Optimization of Photovoltaic Power Systems: Modelization, Simulation and Control* (Springer, Berlin, 2012)
118. N. Maru, J. Vajpai, Model based optimization of tilt angle for solar PV panels in Jodhpur, in *National Conference on Intelligent Systems (NCIS 2014)*, 2014, pp. 32–39
119. P. Koepke, K. Kriebel, B. Dietrich, The effect of surface reflection function and of atmospheric parameters on the shortwave radiation budget. Adv. Space Res. **5**(6), 351–354 (1985)
120. J. Orgill, K. Hollands, Correlation equation for hourly diffuse radiation on a horizontal surface. Sol. Energy **19**(4), 357–359 (1977)
121. A.A. Elbaset, M. Hassan, Design and implement of 100 kW rooftop grid connected PV system: Faculty of engineering as a case study, in *Proceedings of 3rd International Conference on Energy Systems and Technologies*, 2015, pp. 71–83
122. K. Karabacak, N. Cetin, Artificial neural networks for controlling wind–PV power systems: A review. Renew Sustain Energy Rev **29**, 804–827 (2014)
123. A.A. Elbaset, S. Abdelwahab, H. Ibrahim, M. Eid, *Performance Analysis of Photovoltaic Systems with Energy Storage Systems* (Springer, Berlin, 2019)
124. W. Omran, *Performance Analysis of Grid-Connected Photovoltaic Systems*, PhD thesis, University of Waterloo, Canada, 2010
125. N. Ahmed, A. Al-Othman, M. AlRashidi, Development of an efficient utility interactive combined wind/photovoltaic/fuel cell power system with MPPT and DC bus voltage regulation. Electr. Power Syst. Res. **81**(5), 1096–1106 (2011)
126. D. Kumar, K. Chatterjee, A review of conventional and advanced MPPT algorithms for wind energy systems. Renew. Sustain. Energy Rev. **55**, 957–970 (2016)
127. A.A. Elbaset, H. Ali, M.E.A.-E. Sattar, A modified Perturb and observe Algorithm for maximum power point tracking of photovoltaic system using buck–boost converter. J. Eng. Sci. **43**(3), 344–362 (2015)
128. F. Aashoor, F. Robinson, A variable step size perturb and observe algorithm for photovoltaic maximum power point tracking, in *47th International Universities Power Engineering Conference (UPEC)*, IEEE, 2012, pp. 1–6
129. L. Piegari, R. Rizzo, Adaptive perturb and observe algorithm for photovoltaic maximum power point tracking. IET Renew. Power Gener. **4**(4), 317–328 (2010)
130. Y. Yang, F. Blaabjerg, Overview of single-phase grid-connected photovoltaic systems, in *Renewable energy devices and systems*, (CRC Press, Boca Raton, 2017), pp. 41–66
131. A.A. Elbaset et al., Effect of DC link capacitance on stand-alone PV system operation with fluctuated DC resistive loads. Port Said Eng. Res. J. **19**(1), 121–128 (2015)
132. A. Anzalchi, A. Sarwat, Overview of technical specifications for grid-connected photovoltaic systems. Energy Conver. Manage. **152**, 312–327 (2017)
133. N. Femia et al., *Power Electronics and Control Techniques for Maximum Energy Harvesting in Photovoltaic Systems* (CRC Press, Boca Raton, 2012)
134. B. Singh et al., A review of single-phase improved power quality AC-DC converters. IEEE Trans. Ind. Electron. **50**(5), 962–981 (2003)
135. M. Hongbo, F. Quanyuan, A Hybrid modeling and control for buck-boost switching converters, in *International Confererence on Communications, circuits and System*, IEEE, 2009

136. M. Farahat, H. Metwally, A. Mohamed, Optimal choice and design of different topologies of DC–DC converter used in PV systems at different climatic conditions in Egypt. Renew. Energy **43**, 393–402 (2012)
137. O.I. Akihiro, *Design and Simulation of Photovoltaic Water Pumping System*, Master thesis, Faculty of California Polytechnic State University, San Luis Obispo, 2005
138. M. Taghvaee et al., A current and future study on non-isolated DC–DC converters for photovoltaic applications. Renew. Sustain. Energy Rev. **17**, 216–227 (2013)
139. B. Hasaneen, A.A. Elbaset, Design and simulation of DC/DC boost converter, in *2008 12th International Middle-East Power System Conference*, IEEE, 2008, pp. 335–340
140. A.A. Elbaset, H.A. Mohamed, M. Morad, Design and implementation of microcontroller based non-inverting DC/DC buck-boost converter, in *17th International Middle East Power Systems Conference*, Mansoura University, Egypt, 15–17 Dec 2015
141. G.R. Walker, Cascaded DC-DC converter connection of photovoltaic modules. IEEE Trans. Power Electron. **19**(4), 1130–1139 (2004)
142. X. Zhou, T.G. Wang, F.C. Lee, Optimizing design for low voltage DC-DC converters, in *Proceedings of APEC 97 - Applied Power Electronics Conference*, Atlanta, GA, USA, vol. 2, 1997, pp. 612–616
143. S. Kouro et al., Powering the future of industry: High-power adjustable speed drive topologies. IEEE Ind. Appl. Mag. **18**(4), 26–39 (2012)
144. E. da Silva, M.E. Elbuluk, Fundamentals of power electronics, in *Power Electronics for Renewable and Distributed Energy Systems*, (Springer, Berlin, 2013), pp. 7–59
145. A.A. Elbaset, H.H. El-Tamaly Modeling and simulation of photovoltaic/wind hybrid electric power system interconnected with electrical utility, in *2008 12th International Middle-East Power System Conference*, Aswan, pp. 645–649
146. Q.-C. Zhong, T. Hornik, *Control of Power Inverters in Renewable Energy and Smart Grid Integration* (Wiley, Hoboken, 2012)
147. A.A. Elbaset, H. Ali, M.A. El Sattar, Design and performance of single-phase grid inverter photovoltaic system for residential applications with maximum power point tracking, in *Eighteenth International Middle East Power Systems Conference (MEPCON)*, IEEE, 2016, pp. 266–275
148. M. Latran, A. Teke, Investigation of multilevel multifunctional grid connected inverter topologies and control strategies used in photovoltaic systems. Renew. Sustain. Energy Rev. **42**, 361–376 (2015)
149. A. Jhala, R. Gupta, Z-source inverters: design and applications. Int. J. Electron. Commun. Eng. Technol. **6**(12), 17–24 (2015)
150. M. Patil, *Modelling of Grid connected Single Phase Two Stage Full Bridge Converter for Photovoltaic Application*, Master thesis, Oviedo University, 2017
151. S. Maheshri, P. Khampariya, Simulation of single phase SPWM (Unipolar) inverter. Int. J. Innov. Res. Adv. Eng. **1**(9), 12–18 (2014)
152. Y. Sahali, M. Fellah, Regular paper comparison between optimal minimization of total harmonic distortion and harmonic elimination with voltage. J. Electr. Syst. **1**(3), 32–46 (2005)
153. J. Dai, *Current Source Converters for Megawatt Wind Energy Conversion Systems*, PhD thesis, Ryerson University
154. D.G. Holmes, T.A. Lipo, *Pulse Width Modulation for Power Converters: Principles and Practice* (Wiley, Hoboken, 2003)
155. I. Yahyaoui, *Advances in Renewable Energies and Power Technologies: Volume 1: Solar and Wind Energies* (Elsevier, Amsterdam, 2018)
156. M.M. Shaltout, A. Hassan, A. Fathy, Total suspended particles and solar radiation over Cairo and Aswan. Renew. Energy **23**(3–4), 605–619 (2001)
157. A.A. Elbaset, S.I.H. Ata, Optimum photovoltaic solar cells area estimation for feeding micro-grid loads. Int. J. Eng. Res. Technol. **12**(11), 1978–1985 (2019) ISSN 0974-3154
158. L. Huber, M.M. Jovanovic, Single-stage, single-switch, isolated power supply technique with input-current shaping and fast output-voltage regulation for universal input-voltage-range

applications, in *Proceedings of APEC 97-Applied Power Electronics Conference*, vol. 1, 1997, pp. 272–280
159. M.Z. Jacobson, M.Z. Jacobson, *Fundamentals of Atmospheric Modeling* (Cambridge University Press, Cambridge, 2005)
160. D. Kothari, K. Singal, R. Ranjan, *Renewable Energy Sources and Emerging Technologies* (PHI Learning Pvt. Ltd., New Delhi, 2011)
161. B. Babu, K. Mohanty, Doubly-fed induction generator for variable speed wind energy conversion systems-modeling & simulation. Int. J. Comput. Electr. Eng. **2**(1), 1793–8163 (2010)
162. C.E. Dole et al., *Flight Theory and Aerodynamics: A Practical Guide for Operational Safety* (Wiley, Hoboken, 2016)
163. S. Heier, *Grid Integration of Wind Energy: Onshore and Offshore Conversion Systems* (Wiley, Hoboken, 2014)
164. D. Infield, L. Freris, *Renewable Energy in Power Systems* (Wiley, Hoboken, 2020)
165. F.D. Bianchi et al., *Wind Turbine Control Systems: Principles, Modelling and Gain Scheduling Design* (Springer, Berlin, 2006)
166. S. William, Z. Li, *Electricity Generation Using Wind Power* (World Scientific Publishing, Singapore, 2017)
167. A.A. Elbaset et al., *Wind Driven Doubly Fed Induction Generator* (Springer, Berlin, 2018)
168. H. Oyugi, G Mangeni, Wind converters theory, in *Proceedings of Sustainable Research and Innovation Conference*, 2012, pp. 32–36
169. J. Kirsch, *Design of a Small Wind Turbine for Electric Power Generation (1-5kW)*, Thesis, University of Southern Queensland, 2009
170. S.C. Pryor, R. Barthelmie, Climate change impacts on wind energy: A review. Renew. Sustain. Energy Rev. **14**(1), 430–437 (2010)
171. J.F. Manwell et al., *Wind Energy Explained: Theory, Design and Application* (Wiley, Chichester, 2010)
172. M. Ray et al., *Analysis of Wind Shear Models and Trends in Different Terrains* (Renewable Energy Research Laboratory, University of Massachusetts, Amherst, 2006)
173. S. Chowdhury et al., Optimizing the arrangement and the selection of turbines for wind farms subject to varying wind conditions. Renew. Energy **52**, 273–282 (2013)
174. P. Wais, A review of Weibull functions in wind sector. Renew. Sustain. Energy Rev. **70**, 1099–1107 (2017)
175. S.A. Akdağ, A. Dinler, A new method to estimate Weibull parameters for wind energy applications. Energy Conver. Manage. **50**(7), 1761–1766 (2009)
176. B. Wu et al., *Power Conversion and Control of Wind Energy Systems* (Wiley, Chichester, 2011)
177. S. Bhattacharjeea, S. Acharya, PV–wind hybrid power option for a low wind topography, Energy Conversion and Management. Energy Conver. Manage. **89**(1), 942–954 (2015)
178. Q. Van Wyngaardt, *Optimisation of Wind Turbine Electrical Power Conversion*, Master thesis, North-West University, 2011
179. M.R. Altimania, *Modeling of Doubly-Fed Induction Generators Connected to Distribution System Based on eMEGASim® Real-Time Digital Simulator*, Master thesis, University of Tennessee at Chattanooga, 2014
180. Z.M. Dalala et al., Design and analysis of an MPPT technique for small-scale wind energy conversion systems. IEEE Trans. Energy Conver. **28**(3), 756–767 (2013)
181. M. Narayana et al., Generic maximum power point tracking controller for small-scale wind turbines. Renew. Energy **44**, 72–79 (2012)
182. A.A. Albaset, H.H. El-Tamaly, Modeling and simulation of photovoltaic/wind hybrid electric power system interconnected with electrical utility, in *12th International Middle-East Power System Conference*, IEEE, 2008, pp. 645–649
183. W. Cao et al., Wind turbine generator technologies, in *Advances in Wind Power*, (InTech, Rijeka, 2012), pp. 177–204

184. J. Carrasco et al., Power-electronic systems for the grid integration of renewable energy sources: A survey. IEEE Trans. Ind. Electron. **53**(4), 1002–1016 (2006)
185. N.G. Mortensen, et al., Wind atlas for Egypt, in *3rd Middle East-North Africa Renewable Energy Conference (MENAREC 3)*, 2006
186. Catalogue of European Urban Wind Turbine Manufacturers. http://www.urbanwind.net/pdf/CATALOGUE_V2.pdf
187. T. El-Tawil et al., Sizing and rough optimization of a hybrid renewable-based farm in a stand-alone marine context. Renew. Energy **115**, 1134–1143 (2018)
188. R. Banos et al., Optimization methods applied to renewable and sustainable energy: A review. Renew. Sustain. Energy Rev. **15**(4), 1753–1766 (2011)
189. B. Bahmani-Firouzi, R. Azizipanah-Abarghooee, Optimal sizing of battery energy storage for micro-grid operation management using a new improved bat algorithm. Int. J. Electr. Power Energy Syst. **56**, 42–54 (2014)
190. Y. Zhao, et al., The optimal capacity configuration of an independent wind/PV hybrid power supply system based on improved PSO algorithm, in *8th International Conference on Advances in Power System Control, Operation and Management (APSCOM 2009)*, 2009
191. S.A. Pourmousavi et al., Real-time energy management of a stand-alone hybrid wind-microturbine energy system using particle swarm optimization. IEEE Trans. Sustain. Energy **1**(3), 193–201 (2010)
192. Y. Sawle et al., Optimal sizing of standalone PV/Wind/Biomass hybrid energy system using GA and PSO optimization technique. Energy Procedia **117**, 690–698 (2017)
193. L. Sankari, C. Chandrasekar, Semi supervised image segmentation by optimal color seed selection using fast genetic algorithm. Int. J. Comput. Appl. **26**(10), 13–18 (2011)
194. A.A. Elbaset, et al., Design and control strategy of utility interfaced PV/WTG hybrid system, in *The 9th International Middle East Power System Conference, MEPCON*, 2003, pp. 16–18
195. L. Suganthi et al., Applications of fuzzy logic in renewable energy systems – A review. Renew. Sustain. Energy Rev. **48**, 585–607 (2015)
196. A.A. Elbaset et al., Impact of interconnection photovoltaic/wind system with utility on their reliability using a fuzzy scheme. Renew. Energy **31**(15), 2475–2491 (2006)
197. A.M. Eltamaly et al., PSO-based smart grid application for sizing and optimization of hybrid renewable energy systems. PLoS One **11**(8) (2016). https://doi.org/10.1371/journal.pone.0159702
198. A.L. Bukar, C. Tan, A review on stand-alone photovoltaic-wind energy system with fuel cell: system optimization and energy management strategy. J. Clean. Prod. **221**(1), 73–88 (2019)
199. D. Abbes, A. Martinez, G.J.M. Champenois, Life cycle cost, embodied energy and loss of power supply probability for the optimal design of hybrid power systems. Math. Comput. Simul. **98**, 46–62 (2014)
200. A.A. Elbaset, et al., Optimal tuning of PI controller parameters for three-phase AC-DC-AC converter based on Particle Swarm algorithm, in *17th International Middle-East Power System Conference (MEPCON'15)*, 2015
201. M. Taherkhani, R. Safabakhsh, A novel stability-based adaptive inertia weight for particle swarm optimization. Appl. Soft Comput. **38**, 281–295 (2016)
202. C. Ghenai et al., Hybrid solar PV/PEM fuel cell/diesel generator power system for cruise ship: a case study in Stockholm, Sweden. Cases Studied Therm. Eng. **171**(15), 180–189 (2019)
203. C. Ghenai, M.J.E. Bettayeb, Modelling and performance analysis of a stand-alone hybrid solar PV/fuel cell/diesel generator power system for university building. Energy **171**(15), 180–189 (2019)
204. D. Álvaro, R. Arranz, J. Aguado, Sizing and operation of hybrid energy storage systems to perform ramp-rate control in PV power plants. Int. J. Electr. Power Energy Syst. **107**, 589–596 (2019)

Index

A
AC/DC converters, 133–136
Active optical network (AON), 16
Alkaline fuel cell (AFC), 22, 23
Anisotropic models, 66
ANN architecture
  analytic method, 104
  data training, 106, 110
  duty cycle, 102
  input layer, 104
  output layer, 103, 105
  outputs, 109, 113
  regression, 108, 112
  training test, 107, 111
  validation performance, 107, 111
Artificial neural networks (ANN), 48
  analytic method, 52
  data training, 55
  input layer, 57
  output, 59
  output layer, 57
  regression, 58
  training test, 56
  UC sizing, 54
  UC types, 52, 57
  validation performance, 56
Asymmetric DSL (ADSL), 16

B
Base transceiver station (BTS), 7, 11
Battery Storage (BS), 8
Betz Law, 121
Biomass, 5
Buck-boost converter, 75–78
Boost (step-up) converter, 74, 75
Buck (step-down) converter, 73, 74

C
Capacitance, 46
Capacitors, 35
Capacity factor (CF), 128, 129
Carbon-based emissions, 2
Carbon electrode, 44
Cellular networks
  base stations, 12
    architecture, 13
    macrocell, 14
    microcell, 14
  elements, 11
  energy consumption, 12, 13
  energy measurement models, 11
  power consumption, 12, 13
  sectors, 11
Climate data, 81–83, 85
CO_2 emissions, 2, 3
Communication networks, 1
Communication stations, 3
Consumer electronics, 37
Core network, 15
Cost of energy (COE), 7, 161
Current source inverter (CSI), 78
Customer Premises Equipment (CPE), 16

A. A. Elbaset, S. Ata, *Hybrid Renewable Energy Systems for Remote Telecommunication Stations*, https://doi.org/10.1007/978-3-030-66344-5

D
Democratic Republic of Congo, 7
Diesel generators, 1, 19
Differential capacitance, 39
Diffuse irradiance, 66
Direct methanol FC (DMFC), 22–23
Direct solar radiation, 63
Discrete regenerative fuel cell (DRFC), 25

E
Economic and political conditions, 118
Efficiency, 119
Electrical load, 164
Electric power generation, 117
Electrochemical capacitors, 36
Electrolyzer (EZ), 24
Electrostatic capacitor, 35
Energy, 2
Energy balance, 67, 86, 87, 130, 131
Energy Management System (EMS), 6
Energy system, 3
Equivalent series resistance (ESR), 43
Equivalent series resistor (ESR), 40
Evolutionary algorithm (EA), 160

F
Fiber-To-The-Building (FTTB), 16
Fiber-To-The-Curb (FTTC), 16
Fiber-To-The-Home (FTTH), 16
Fiber-To-The-Node (FTTN), 16
Fossil fuels, 1, 2
Frequency-dependent effects, 46
Fuzzy logic control (FLC) techniques, 69

G
Gas diffusion electrode (GDE), 23
Geographic factor method, 63
Geothermal energy, 5
Gigabit PON (GPON), 16
Greenhouse gas, 2
Ground reflection radiation, 65

H
Heat radiation, 2
High-switching frequency, 79
Hybrid Optimization Model for Electric Renewables (HOMER), 7, 8
Hybrid Solar/Wind Power system (HSWPS), 8
Hybrid system
  controllers, 163
  converters, 163
  energy management process, 163, 164
  load demand model, 162
  operation strategy, 163, 164
  RFC, 163
  solar panels, 162
  UC, 163
  wind turbine, 162, 163
Hydraulics, 5
Hydrogen balance, 32
Hydrogen production
  angle, 88
  computer program, 94
  H-3000, 93
  HyLYZER^{TM}-2, 93
  latitude angle, 88
  Marsa Alam, 97
  Minia, 96, 98, 99
  regenerative fuel cells system, 88
  Shark El-Oweinat, 95, 97, 98
  solar panel orientation, 88
  surplus/deficit, 95
  usage, 96
  water tanks, 99
Hydrogen storage, 31

I
Impedance source inverter, 78
Incremental conductance (INC) technique, 69
Information and communication technology (ICT), 2
Information technology (IT), 2
Initial Capital, 7
Inverters
  CSI, 78
  frequency, 78
  impedance source inverter, 78
  magnitude, 78
  multi-level, 79
  output voltage, 78
  single-phase bridge inverters, 78, 79
  sinusoidal, 78
  three-phase full-bridge VSI, 80
  VSI, 78

K
Kinetic energy, 118

L
Levelized energy cost (LEC), 164, 172
Loss of power supply probability (LPSP), 8

M
Maximum power point tracking (MPPT)
  converter acting, 69
  electronic device, 68
  NN, 69
  photovoltaic generation, 68
  PV power systems, 68
  solar panel outputs power, 68
Metro network, 15
Microbial fuel cell (MFC), 23
Mobile switching centers (MSCs), 11
Mobile terminals, 11
Molten carbonate FC (MCFC), 22
Multi-level inverters, 79

N
Network technologies, 2
Neural networks (NN), 69
Nominal Operating Cell Temperature (NOCT), 67
Nuclear power, 2

O
On-grid photovoltaic power systems, *see* Photovoltaic power systems (PVPS)
Optical Line Termination (OLT), 17
Optimum solar cells area (OSCA), 67, 81
Oxygen balance, 32
Oxygen production, 29

P
Particle Swarm Optimization technique (PSO), 161, 168
Passive optical network (PON), 16
Permanent magnet (PM) generator, 133
Permanent magnet synchronous generator (PMSG), 133, 134
Perturbation and observation (P&O) technique, 69
Phosphoric acid electrolyte FC (PAFC), 22
Photovoltaic power systems (PVPS), 8
  angle, 88
  azimuth, 62
  climate data, 81, 83, 85
  communication stations, 81
  components, 63
  control grid, 80
  daily beam radiation, 65
  DC-DC converters
    boost (step-up) converter, 74, 75
    buck-boost converter, 75–78
    buck (step-down) converter, 73, 74
    electrical drives, 72
    function, 72
    switching-mode regulators, 71
  double-stage, 70
  Earth surface, 61
  energy balance programs, 86, 87
  full-bridge (FB) inverter, 71
  geographic factor method, 63
  grid-connected PV systems, 70
  ground reflection radiation, 65
  half-bridge (HB) inverter, 71
  latitudes, 81, 88
  LCL filter, 71, 72
  longitudes, 81
  maximum power MPPT
    OSCA, 67
    solar cells, 68
  OSCAs, 82
  parameters, 61, 63, 64
  radiation, 81–83, 85
  single-stage, 70
  sky diffused radiation, 66
  solar energy, 61
  solar panels, 61, 81, 86, 95, 96, 101
  solar PV array, 62
  solar radiation, 61
  stand-alone PV/FC/UC, 62
  surplus/deficit energy, 89–93
  techniques, 63
  topology, 71
  wind energy, 81
Photovoltaic systems, 5
Physical layer technologies, 15
Point to multipoint networks (P-t-MP), 16
Point to point (P-t-P), 16
Polymer electrolyte membrane FC (PEMFC), 21, 22
Power coefficient (C_p), 121, 122, 157
Power electronic circuit, 135
Power extraction, 120
Protonic ceramic fuel cell (PCFC), 23
PSO optimization algorithm
  acceleration constants, 168
  ALC, 177, 178
  artificial intelligence, 167
  economic optimization, 172
  electrical demand, 173
  local optimum solution, 168
  Marsa Alam, 177
  MATLAB environment, 173

PSO optimization algorithm (*cont.*)
Minia, 177
national electric grid, 173
optimal solution, 169
parameters, 169
particles, 167, 168
process, 173
Shark El-Oweinat, 177
social behavior, 167
UC, 173
Pulse-width modulation (PWM) techniques, 79, 80
Pyranometer, 66

R
Radiation, 81–83, 85
Radio base stations (RBSs), 11
RC serial model, 40, 41
Rectifier circuits, 133
Regenerative fuel cells (RFC), 163
AFCs, 22, 23
air exit flow rate, 30
air usage, 30
applications, 19
batteries, 19
communication network infrastructures, 19
diesel generators, 19
DMFCs, 22
DRFC, 31, 32
efficiency, 26, 27
electrochemical reaction, 19
energy conversion device, 19
hydrogen/hydrogen-rich fuels, 19
hydrogen production, 27–29
hydrogen storage, 31
hydrogen usage, 29, 30
inputs, 20
MCFC, 22
MFCs, 24
outputs, 20
oxygen production, 29
PAFC, 22
PCFCs, 23
PEMFCs, 21, 22
principles, 21, 24–26
SOFC, 22
telecom backup power solutions, 19
telecommunication, 20
ZAFC, 23
Remote areas, 1, 9
Remote node (RN), 16
Renewable energy base stations (REBS), 8
Renewable energy systems (RES)
advantages, 159
Aircon wind turbine, 180
Aircon WTG type, 171
approach, 9
artificial intelligence algorithms, 160
base station power consumption, 179
batteries, 6, 179
biomass, 5
communication stations, 9, 180
cost of energy, 167
DC-DC converters, 179
diesel machine, 179
economic and technical parameters, 169, 170
electric load, 161
energy management, 8
energy systems, 6
environmental variations, 7
fuel cells, 6, 9
Gaia wind turbine, 180
Gaia WTG type, 170
geothermal energy, 5
hybrid energy harvesting systems, 6
hybrid power generation, 159
hybrid power systems, 7
hybrid PV/WT/RFC/UC power system, 160
hybrid PV/WT/RFC/UC/UG power system, 161
hybrid system, 10, 159
hydraulics, 5
load profile, 169
Marsa Alam Site, 171, 175, 180
MATLAB software, 161
microgrid, 5
Minia, 176, 180
optimal solution, 161
optimization techniques, 160
optimum design, 6
OSCA, 180
PELV, 169
PELW, 169
photovoltaic (PV) array, 6, 8
photovoltaic (PV) power system, 9
pollution emissions, 159
power communication stations, 5
PSO algorithm, 160
Shark El-Oweinat site, 169, 174
sizing optimization, 8
solar energy, 4, 5
solar panels, 160, 170, 171
solar PV, 159
SPR-X22-360 solar panel, 180

supply communication stations, 179
techno-economic analysis, 6
technological feasibility, 8
telecommunication base stations, 6
TPC, 167
UC, 9
wind energy, 4, 10
wind–PV, 6
wind turbines, 8, 159, 160

S
Self-excited asynchronous generator (SEIG), 134
Single-phase bridge inverters, 78, 79
Single-phase rectifiers, 133
Sinusoidal pulse-width modulation (SPWM), 79
Sky diffused radiation, 66
Smart Grid, 4
Soft computing method, 69
Solar cells, 68
Solar energy, 4, 5, 61, 115
Solar panels, 61, 81, 86, 95, 96, 101, 162
Solar radiation, 1, 61, 63, 64, 66, 112
Solid oxide FC (SOFC), 22
Stand-alone system, 1
State of charge (SOC), 161
System cost analysis
cost estimation
discrete regenerative fuel cells cost, 165, 166
economic analysis, 164
electrolyzer part cost, 166
photovoltaic panel cost, 165
power conditioner and wiring, 166
UC, 166
wind turbine cost, 165
problem statement, 164

T
Telecom networks, 3
Telecom operators, 1
Telecommunication, 2, 3
Temperature correction factor (TCF), 38
Temperature dependency, 38
Three-phase bridge rectifier, 133
Three-phase full-bridge VSI, 80
Total harmonic distortion (THD), 134
Total net present cost (TNPCC), 7, 161
Total present cost (TPC), 167, 172
Traditional grid infrastructure, 4
Transportation industry, 36

U
Ultracapacitor (UC), 163
accurate sizing
capacitance, 49
discharge duration difference, 53
duration of discharge, 51–53
percentage error, 53, 54
temperature change, 49
temperature effect, 47, 48
time difference, 52
voltage change, 49
voltage effect, 44, 46, 47
ANN, 48
batteries, 35
capacitance, 35
capacitor, 35
cell construction, 36
characteristics
stacking, 39
temperature dependency, 38
voltage dependency, 38, 39
computer program flowchart, 49
consumer electronics, 37
critical-load operations, 37
duration of discharge, 51
electric components, 35
electrical models
first-order model, 41
RC serial model, 40, 41
electricity, 35
K2 series, 50
lead-acid battery, 37
photovoltaic panels/wind, 35
simple sizing, 42–45, 50
storage element, 35
telecommunication devices and stations, 48
transportation, 36
UPS, 37
Uninterrupted power supply (UPS) systems, 37
Unitized regenerative fuel cell (URFC), 25, 26

V
Voltage dependency, 38, 39
Voltage effect, 44, 46, 47
Voltage source converter (VSC), 135
Voltage source inverter (VSI), 78

W
Water balance, 32
Weibull distribution, 125, 126
Wind converter system, 123

Wind energy system, 2, 4
 AC/DC converters, 133–136
 ANN architecture, 151, 153, 154, 156, 157
 ANWTG, 137
 CF, 137
 communication station, 136
 conversion system, 132
 design algorithms
  ANWTG, 129
  CF, 129
  energy balance, 130, 131
  optimum number, 130, 131
  Weibull parameters, 129
  wind velocities, 129
 electrical generator, 133
 electrical power, 132
 energy balance, 140
 grid-connected inverter structure, 133
 grid-connected WECS, 132
 heterogeneity, 115
 Matlab functions, 136
 power electronic converter topology, 133
 power electronic energy conversion system, 133
 solar energy, 115
 stand-alone, 132
 water and territory surfaces, 115
 Weibull parameters, 136, 137
 wind turbine (*see* Wind turbine)
 WTGs, 137
Wind generation, 117
Wind machine, 119
Wind parameter
 classes, 123, 124
 power output, 126, 127
 shear, 124
 speed, 123
 Weibull distribution, 125, 126
Wind-powered electricity generation, 122, 123
Wind shear, 124
Wind speed, 1
Wind turbine generator (WTG)
 ANWTG, 128
 average power output, 128
 capacity factors, 139, 140
 CF, 128
 DC-DC converter, 146
 Duty ratio *vs.* output current, 150
 energy production, 128
 Hydrogen prod, 145–149
 hydrogen surplus/defect, 148, 149
 Marsa Alam, 138
 Minia Site, 137
 model of equation, 128
 ONWTG, 140
 surplus/deficit energy, 141, 143–145
 WES, 151
 wind energy system, 128
Wind turbines
 components, 115
  base, 117
  basic elements, 116
  brake, 117
  controller, 117
  gearboxes and direct drives, 117
  generator, 117
  nacelle, 117
  rotor blades, 117
  tower, 117
  yaw mechanism, 117
 efficiency considerations, 122, 123
 electrical energy, 115
 electrical power, 115
 kinetic energy, 115
 power coefficient, 121
 power output, 117–119
 pressure gradient force, 116
 rotary mechanical power, 115
 theoretical power, 119, 120
Wind velocity, 120
Wired access network, 15
Wired communication networks, 15
Wired networks energy consumption, 14–17
Wireless cellular network power consumption, 12
Wireless network, 3

Z
Zinc anode, 23
Zinc-air fuel cell (ZAFC), 23